AF581861

Gas-Liquid Two-Phase Flow in the Pipe or Channel

Gas-Liquid Two-Phase Flow in the Pipe or Channel

Editors

Maksim Pakhomov
Pavel Lobanov

MDPI • Basel • Beijing • Wuhan • Barcelona • Belgrade • Manchester • Tokyo • Cluj • Tianjin

Editors
Maksim Pakhomov
Russian Academy of Sciences
Russia

Pavel Lobanov
Russian Academy of Sciences
Russia

Editorial Office
MDPI
St. Alban-Anlage 66
4052 Basel, Switzerland

This is a reprint of articles from the Special Issue published online in the open access journal *Water* (ISSN 2073-4441) (available at: https://www.mdpi.com/journal/water/special_issues/gas_liquid_two_phase_flow).

For citation purposes, cite each article independently as indicated on the article page online and as indicated below:

LastName, A.A.; LastName, B.B.; LastName, C.C. Article Title. *Journal Name* **Year**, *Volume Number*, Page Range.

ISBN 978-3-0365-3387-2 (Hbk)
ISBN 978-3-0365-3388-9 (PDF)

Contents

About the Editors

Pakhomov Maksim (PhD, Dr. of Sciences, Professor, Leading Researcher) was born in 1976 in the city of Novosibirsk and graduated from Novosibirsk Technical State University in 1999. From 1997 to the present, he has been an employee of the Institute of Thermophysics SB RAS (Novosibirsk).

His research interests include the CFD of hydrodynamics, as well as the heat and mass transfer in turbulent two-phase droplet-laden and bubbly flows. Other topics of his research include flow and heat transfer in a turbulent, non-Newtonian, Bingham-Schwedoff fluid and synthetic impinging jets. He is the author of more than 200 publications, including 3 monographs and 30 articles in international journals.

Lobanov Pavel (PhD, Senior Researcher) was born in 1980 and graduated from Novosibirsk Technical State University in 2002. From 2000 to the present, he has been an employee of the Institute of Thermophysics SB RAS (Novosibirsk).

Lobanov Pavel is a specialist in hydrodynamics and thermophysics. The main directions of his research are: hydrodynamics, heat and mass transfer and the turbulent structure of gas–liquid two-phase flows; nuclear reactor thermal-hydraulics; study of processes of thermal fatigue failure in nuclear power plants; and fluid–structure interaction (FSI), studied on the base of a coolant cross-flow in a system of two consecutive cylinders and a longitudinal flow in the annular channel. Water and heavy liquid-metal coolants were used. The database containing these measurements is used in different laboratories to develop and verify techniques for flow prediction such as Euclid, Hydra/IBRAE/LM, Conv-3d, KABARE, LOGOS, NEK5000, and approaches based on commercial CFD codes.

MDPI

Editorial

Gas-Liquid Two-Phase Flow in a Pipe or Channel

Maksim A. Pakhomov [1,*] and Pavel D. Lobanov [2]

[1] Laboratory of Thermal and Gas Dynamics, Kutateladze Institute of Thermophysics, Siberian Branch of Russian Academy of Sciences, Acad. Lavrent'ev Avenue, 1, 630090 Novosibirsk, Russia

[2] Laboratory of Problems of Heat and Mass Transfer, Kutateladze Institute of Thermophysics, Siberian Branch of Russian Academy of Sciences, Acad. Lavrent'ev Avenue, 1, 630090 Novosibirsk, Russia; lobanov@itp.nsc.ru

* Correspondence: pakhomov@ngs.ru or pakhomov@itp.nsc.ru

Citation: Pakhomov, M.A.; Lobanov, P.D. Gas-Liquid Two-Phase Flow in a Pipe or Channel. *Water* **2021**, *13*, 3382. https://doi.org/10.3390/w13233382

Received: 29 October 2021
Accepted: 11 November 2021
Published: 1 December 2021

This Special Issue contributes to highlight and discusses topics related to various aspects of the two-phase gas-liquid flows. They can be used both in fundamental sciences and practical applications. We consider that the main goal has been successfully reached. This Special Issue received investigations from Russia, China, Thailand, ROC-Taiwan, Saudi Arabia, and Pakistan. We are happy to see that all papers present findings characterized as unconventional, innovative, and methodologically new. We hope that the readers of the journal Water can enjoy and learn about experimental and numerical study of two-phase flows using the published material, and share the results with the scientific community, policymakers and stakeholders.

Two-phase gas-liquid flows are frequently applied in energy, nuclear, chemical, geothermal, oil and gas and refrigeration industries. Two-phase gas-liquid flows can occur in various forms, such as flows transitioning from pure liquid to vapor as a result of external heating, separated flows, and dispersed two-phase flows where one phase is present in the form of droplets, or bubbles (i.e., liquid or gas) in a continuous carrier fluid phase (i.e., gas or liquid). Typically, such flows are turbulent with a considerable interfacial interaction between the carrier fluid and the dispersed phases. The variety of flow regimes complicates significantly the theoretical prediction of hydrodynamics of the two-phase flow. It requires application of numerous hypotheses, assumptions, and approximations. Often the complexity of flow structure allows theoretical description of its behavior, and so empirical data are applied instead. The correct simulation of two-phase gas-liquid flows is of great importance for safety and design of energy equipment elements.

The simultaneous measurements of the hydrodynamic and thermal characteristics of spray cooling were performed in [1]. The size of individual droplets before their impact on the forming liquid film and estimation of the number of droplets falling onto the impact surface from the impinging spray was performed using the high-speed recording with high spatial resolution. The authors showed various possible scenarios for this interaction, such as the formation of small-scale capillary waves during impacts of small droplets, the appearance of "craters", and splashing crowns in the case of large ones. Evolution of the non-steady state temperature fields during spray cooling in regimes without boiling was measures using high-speed infrared thermography. It was shown that, for the studied regimes, the heat transfer weakly depends on the heat flux density and is primarily determined by the mass flow rate of the spray.

The flow structure, turbulence modification, and heat transfer augmentation in a droplet-laden dilute flow over a single-side backward-facing step were numerically studied in [2]. Numerical simulations were performed for water droplets, with inlet droplet diameters d_1 = 1–100 μm and the mass fraction M_{L1} = 0–0.1. There was almost no influence of a small number of droplets on the mean gas flow and coefficient of wall friction. A substantial heat transfer augmentation in a droplet-laden mist separated flow was shown. Heat transfer enhancement was revealed both in the flow recirculation and flow relaxation zones for fine dispersed droplets, and the largest droplets augmented heat transfer mainly after the flow reattachment point. The largest heat transfer enhancement in a droplet-laden flow was obtained for small evaporating particles.

Authors of [3] carried out the experimental and numerical study of the mean and turbulent flow structure and heat transfer in a bubbly gas-liquid flow at a single-side backward-facing step. The flow was directed upward in a duct. The carrier fluid phase velocity was measured using the PIV/PLIF system. The set of two-dimensional RANS equations was used for modelling two-phase bubbly flow. The motion and heat transfer in the dispersed phase were modelled using the Eulerian approach with taking into account the bubbles break-up and coalescence. The method of delta-functions was employed to simulate the distribution of polydispersed gas bubbles. Small bubbles were presented over the entire duct cross-section and the larger bubbles were mainly observed in the shear mixing layer and flow core. The recirculation length in two-phase bubbly flow was shorter (almost twice) than in the case of single-phase fluid flow. The position of the heat transfer maximum was located after the reattachment point. The addition of air bubbles leaded to a significant increase in heat transfer (up to 75%).

An overview of the methods for predicting the pressure drops and heat transfer of two-phase flows in small-diameter ducts and a comparison of several methods were presented in [4]. The comparison was performed for the conditions of high reduced pressures $p_r = p/p_{cr} \approx 0.4$–$0.6$, where p_{cr} is the critical pressure. The results of authors experimental studies of pressure drops and flow boiling heat transfer of freons in the region of low and high mass flow rates $G = 200$–$2000\ \mathrm{kg/(m^2 \cdot s)}$ were presented in the paper. A description of the experimental stand was given, and a comparison of own experimental data with those obtained using the most reliable calculated relations was carried out.

The experimental study of the flow structure and the water holdup in an oil–water–gas three-phase flow in a horizontal and vertical pipe was carried out in [5]. The three-phase flow consisted of white mineral oil, distilled water, and air. The flow pattern maps in terms of the Reynolds number and the ratio of the superficial velocity of the gas to that of the liquid mixture for different Froude numbers were given by authors. The relationship between the transient water holdup and the changes of the flow patterns in horizontal and vertical sections of the pipe were presented. Authors presented the dimensionless power-law correlation for the water holdup in the vertical section.

The numerical and experimental study of ultrasonic atomization of fluids using silicon-based three Fourier-horn ultrasonic nozzles was carried out in [6]. COMSOL Multiphysics 5.4 was used to perform numerical analysis. The design and characteristics of microprocessor-based ultrasonic nozzles based on silicon with a frequency of 485 kHz were presented. During operation, deionized water was initially sprayed onto the microdroplets, which were stably and continuously formed. A new ultrasonic nebulizer promotes the development of resonance of capillary surface waves at a given frequency. The developed device is compact and energy-saving. It can be used in the green energy industry and non-invasive drug delivery.

Poincare-Light Hill Technique was used for the theoretical analysis of wall shear stress on the MHD two-phase fluctuating flow of dusty fluid in [7]. The flow between two parallel non-conducting plates was considered. The conversion of heat created a fluid flow. Spherical dust particles were evenly dispersed in the base fluid. It was pointed that an increase in Grashof number, radiation variable parameter, and dusty parameter led to an increase in velocities of carrier fluid and dusty particles. On the other hand, a decrease in phase velocities was found with an increase in magnetic parameter. An increase in radiation caused an increase in temperature. The behavior of base fluid and dusty particles was similar.

Authors of [8] studied the non-chemical treatment of aqueous systems with humid air exposed to IR waves. It was shown for the first time that in the samples of deionized water and mineral water, the redox potential and surface tension decreased, and the dielectric constant increased. The treatment of carbonate or phosphate buffers leads to a significant increase in their buffering capacity against acidification and leaching. In water samples treated with humid air without IR processing, no changes were observed.

New experiments have been carried out to study the characteristics of the development of the self-aeration region in a flow along a flat chute [9]. A double-tip conductivity probe was used. The self-aerating area with a free surface was from 27.16% to 51.85% of the liquid depth. The average and fluctuational velocity of the flow in the transverse direction increased downstream the flow inlet. In this area, fluctuations in the flow velocity in the cross-sections were smoothed out as the flow develops. Higher velocity fluctuations in the direction corresponded to the presence of much stronger turbulence increased the diffusion of air bubbles from the free surface of the water to the flow.

Last but not least, we want to thank Aroa Wang—The Assistant Editor, MDPI—For her dedication and willingness to publish this special issue. She has been a major supporter of the special issues, and we are indebted to her.

Author Contributions: Conceptualization, M.A.P. and P.D.L.; methodology, M.A.P. and P.D.L.; writing—original draft preparation, M.A.P. and P.D.L.; writing—review and editing, M.A.P. and P.D.L.; resources, M.A.P. and P.D.L.; project administration, M.A.P. and P.D.L.; All authors have read and agreed to the published version of the manuscript.

Funding: This research received no external funding.

Institutional Review Board Statement: Not applicable.

Informed Consent Statement: Not applicable.

Conflicts of Interest: The authors declare no conflict of interest.

References

1. Serdyukov, V.; Miskiv, N.; Surtaev, A. The simultaneous analysis of droplets' impacts and heat transfer during water spray cooling using a transparent heater. *Water* **2021**, *13*, 2730. [CrossRef]
2. Pakhomov, M.; Terekhov, V. Droplet evaporation in a two-phase mist turbulent flow behind a backward-facing step. *Water* **2021**, *13*, 2333. [CrossRef]
3. Bogatko, T.; Chinak, A.; Kulikov, D.; Lobanov, P.; Pakhomov, M. The effect of a backward-facing step on flow and heat transfer in a polydispersed upward bubbly duct flow. *Water* **2021**, *13*, 2318. [CrossRef]
4. Belyaev, A.V.; Dedov, A.V.; Krapivin, I.I.; Varava, A.N.; Jiang, P.; Xu, R. Study of pressure drops and heat transfer of nonequilibrial two-phase flows. *Water* **2021**, *13*, 2275. [CrossRef]
5. Ren, G.; Ge, D.; Li, P.; Chen, X.; Zhang, X.; Lu, X.; Sun, K.; Fang, R.; Mi, L.; Su, F. The flow pattern transition and water holdup of gas–liquid flow in the horizontal and vertical sections of a continuous transportation pipe. *Water* **2021**, *13*, 2077. [CrossRef]
6. Song, Y.-L.; Cheng, C.-H.; Reddy, M.K. Numerical analysis of ultrasonic nebulizer for onset amplitude of vibration with atomization experimental results. *Water* **2021**, *13*, 1972. [CrossRef]
7. Khan, D.; Rahman, A.U.; Ali, G.; Kumam, P.; Kaewkhao, A.; Khan, I. The effect of wall shear stress on two phase fluctuating flow of dusty fluids by using light hill technique. *Water* **2021**, *13*, 1587. [CrossRef]
8. Yablonskaya, O.; Voeikov, V.; Buravleva, E.; Trofimov, A.; Novikov, K. Physicochemical effects of humid air treated with infrared radiation on aqueous solutions. *Water* **2021**, *13*, 1370. [CrossRef]
9. Song, L.; Deng, J.; Wei, W. Air diffusion and velocity characteristics of self-aerated developing region in flat chute flows. *Water* **2021**, *13*, 840. [CrossRef]

 water

Article

The Simultaneous Analysis of Droplets' Impacts and Heat Transfer during Water Spray Cooling Using a Transparent Heater

Vladimir Serdyukov *, Nikolay Miskiv and Anton Surtaev

Laboratory of Low Temperature Thermophysics, Kutateladze Institute of Thermophysics SB RAS, Academician Lavrent'ev Avenue 1, 630090 Novosibirsk, Russia; nikerx@gmail.com (N.M.); surtaev@itp.nsc.ru (A.S.)

* Correspondence: vsserd@gmail.com

Abstract: This paper demonstrates the advantages and prospects of transparent design of the heating surface for the simultaneous study of the hydrodynamic and thermal characteristics of spray cooling. It was shown that the high-speed recording from the reverse side of such heater allows to identify individual droplets before their impact on the forming liquid film, which makes it possible to measure their sizes with high spatial resolution. In addition, such format enables one to estimate the number of droplets falling onto the impact surface and to study the features of the interface evolution during the droplets' impacts. In particular, the experiments showed various possible scenarios for this interaction, such as the formation of small-scale capillary waves during impacts of small droplets, as well as the appearance of "craters" and splashing crowns in the case of large ones. Moreover, the unsteady temperature field during spray cooling in regimes without boiling was investigated using high-speed infrared thermography. Based on the obtained data, the intensity of heat transfer during spray cooling for various liquid flow rates and heat fluxes was analyzed. It was shown that, for the studied regimes, the heat transfer coefficient weakly depends on the heat flux density and is primarily determined by the flow rate. In addition, the comparison of the processes of spray cooling and nucleate boiling was made, and an analogy was shown in the mechanisms that determine their intensity of heat transfer.

Keywords: spray cooling; transparent heater; high-speed video recording; infrared thermography

Citation: Serdyukov, V.; Miskiv, N.; Surtaev, A. The Simultaneous Analysis of Droplets' Impacts and Heat Transfer during Water Spray Cooling Using a Transparent Heater. *Water* **2021**, *13*, 2730. https://doi.org/10.3390/w13192730

Academic Editor: Maksim Pakhomov and Pavel Lobanov

Received: 29 July 2021
Accepted: 28 September 2021
Published: 2 October 2021

1. Introduction

Today, spray cooling is one of the most effective, reliable, and demanded modes of cooling and thermal stabilization of various heat-generating devices. Among its main advantages are the ability to remove high heat fluxes while ensuring a sufficiently uniform surface temperature, the ability to cool objects of complex (not flat) geometry and with a relatively large area even using a single nozzle, and a low liquid flow rate [1–4]. For these reasons, this process nowadays is used in different industrial cycles and technologies. In addition to traditional quenching in metallurgy, spray cooling is used in medical and aerospace technologies, as well as in fire safety systems. Moreover, application and optimization of jet and spray cooling systems are widely discussed today in relation to the development of high heat flux cooling solutions [5], e.g., heat exchangers for hydrogen storage, rocket nozzles, laser and microwave directed energy weapons, advanced radars, and so on. Finally, the relevance of the development of such systems for cooling high-performance microelectronic devices [6], computers and data centers [7], and high-power LEDs [8] should be noted. As the literature analysis shows, in the majority of applications with spray cooling, the main working fluid is water, first of all, because of its high heat capacity and enthalpy of vaporization, which makes it possible to remove sufficiently high heat fluxes up to 900 W/cm^2 [9]. In addition, this coolant has a low cost and high environmental friendliness, and is simple and safe to use.

In this regard, many studies are devoted today to the processes during spray cooling, the description of which could be found in recent review papers [3,4,10]. At the same time, the analysis of these works shows that some issues related to the dynamics of liquid irrigation and heat transfer during spray cooling still remain open. First of all, this is because of the fact that spray cooling is an extremely difficult process for experimental study, and it is necessary to analyze a whole complex of its hydrodynamic and thermal characteristics. Often, it is possible to investigate in detail only one or several parameters of spray irrigation using a certain measuring technique. In particular, one of the most important parameters is the size and velocity of droplets in the gas-droplet flows. Today, these characteristics are measured using such methods as a side-view video recording, laser phase-Doppler particle analyzer, laser diffraction, interferometric particle imaging, and so on [11]. However, these methods do not allow to analyze the patterns and features of spray irrigation.

There are also many questions related to the heat transfer during spray cooling. In the overwhelming majority of experimental studies, measurements of thermal characteristics are carried out using thermocouples and local temperature sensors with low temporal and spatial resolutions. At the same time, as the results of a number of works show, the temperature field of a heater during spray cooling can be non-uniform depending on the irrigation mode [12]. General data do not allow to obtain information on the local unsteady characteristics of heat transfer and to assess their contribution to integral heat transfer. This hindered the understanding of the governing mechanisms of heat transfer, which is very important for the development of theoretical models to describe the process and engineering methods to calculate systems with jet or spray cooling. This indicates the need to introduce new methods with high spatial and temporal resolutions for the comprehensive study of the spray and jet irrigations' features.

As sprays and aerosols represent particle-laden flows [13,14], some questions related to the flow dynamics and features of droplets motion are also of interest. For example, the clustering and collision of droplets can affect how they will interact with an impact surface. In turn, this can have a significant influence on the local and, consequently, total heat transfer rate during spray cooling. Today, in the literature, numerous papers devoted to both the experimental and numerical investigations of the mentioned and other phenomena could be found. For example, the authors of [15,16] studied the relative motion of monodisperse high Stokes number particle pairs. The effect of gravity on the clustering and collision of bidisperse inertial particles was investigated in [17]. In [18], the relative dispersion of tracer particles was analyzed.

To study the processes of heat and mass transfer in various two-phase systems, the transparent design of a heating surface is actively used today. For example, such a design has been successfully applied by a number of researchers to analyze the characteristics of liquid boiling using high-speed video recording from the bottom side of a heating surface [19–23]. An analysis of the presented results shows that this makes it possible to study in detail the dynamics of vapor bubbles, to carry out a detailed statistical analysis of the bubbles departure diameters, the emission frequencies, and the nucleation site density in a wide range of heat loads. Moreover, this recording format allows to study the dynamics of the triple contact line at the bubble base under various conditions [24–26].

In addition to the visualization of two-phase systems' dynamics, the design of a transparent heating element also makes it possible to simultaneously study the surface unsteady temperature field using high-speed thermographic recording. For example, the authors of [27–31] used heaters based on a sapphire substrate with a deposited transparent indium-tin oxide film. As the result, both integral surface temperature and the features of local heat transfer in the vicinity of a triple contact line during nucleate boiling of various liquids were analyzed. In addition, such a recording was used to study the heat transfer during droplet impact on a liquid film. In particular, in [32], a water droplet impact on a thin heated wafer, which is being cooled by a film flow generated from water jet impingement, was studied using IR recording. The authors showed that the drop impact breaks the steady-state cooling and causes a change in local temperature around the drop landing location.

The usage of a transparent heating surface also showed its perspectives for studying the interface evolution during the interaction of a single droplet and liquid film [33–36]. In particular, the authors of [33] showed various scenarios for drop impact on a film of water and a glycerol–water mixture. However, studies using such a promising design of the impact surface to investigate the characteristics of spray or multi-jet cooling today are still of a single nature. Here, only an early paper can be noted, for example, refs. [37–39] and others, in which the authors used transparent surfaces primarily to study integral irrigation patterns and identify the interaction between bubbles and impinging droplets. However, issues such as the sizes and number of droplets falling onto the surface were not considered. In addition, there are few works devoted to the study of the surface temperature distribution during spray cooling. It is obvious that carrying out experimental studies devoted to a detailed consideration of all key aspects of the mass and heat transfer will make it possible to obtain a large array of new experimental information. Such information will be useful for the creation of generalized dependencies describing the intensity of spray cooling.

The aim of the present study was the comprehensive investigation of the spray cooling features using a special design of the impact surface. Experimental data on the main hydrodynamic and thermal characteristics of spray irrigation were obtained using high-speed video visualization and thermography from the reverse side of a transparent heating element.

2. Materials and Methods

2.1. Experimental Setup and Transparent Heater Construction

The principal scheme of the experimental setup used to study the hydrodynamic and thermal parameters of spray cooling is presented in Figure 1. The setup consists of four basic sections: a circuit to supply gas-droplet multiphase flow, a spray nozzle, a transparent impact surface with a heating system, and high-speed video/infrared cameras.

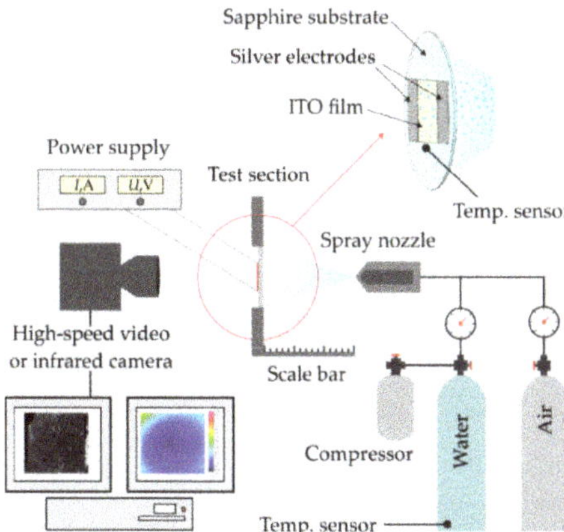

Figure 1. Schematic view of the experimental setup and the test section.

The key feature of the experimental setup is an optically transparent impact surface with a thin-film heater based on indium tin oxide (ITO), deposited onto a sapphire substrate by ion sputtering (ISP SB RAS). In this method, a commercially fabricated ITO target (Girmet Ltd., Moscow, Russia) was bombarded with argon ions. Such a technique allows one to fabricate smooth and uniform films with a given thickness. The main property of ITO is transparency in the visible range of wavelengths (380–750 nm) and opacity in the IR range (3–5 μm), which makes it possible to simultaneously record the irrigation pattern using video recording and measure the surface temperature using IR thermography. A similar heater design was previously used in experiments on liquid boiling [25–27] and has shown its high prospect for the simultaneous study of the evolution of the liquid–vapor system and the temperature field of the heating surface. The thickness of the used sapphire substrate was 400 μm and the thickness of the ITO film heater was 1 μm. The area of direct heat release was 20 × 20 mm^2 and the spray irrigated the vertically oriented sapphire surface. The density of the supplied heat flux was measured according to the current passed through the heater and the potential difference between the silver current leads, also deposited by the ion sputtering technique along the edges of the ITO heater. At this stage of the study, the single-phase heat transfer during spray cooling without the development of boiling was studied, and the maximum heat flux density was 133 kW/m^2.

2.2. Spray Flow Parameters

The BKT-Engineering SS 4230 nozzle made of stainless steel and providing full cone spray was used in the experiments. The orifice diameter of the selected nozzle was 1.7 mm. Deionized MilliQ water was used as a working fluid and air was used as a working gas. The temperature of the working phases in all experiments was equal to room temperature (25 °C) and the relative humidity of the ambient air was 65%. The nozzle-to-surface distance was configured such that the spray impact area just inscribed the heating surface. Thus, to define the mean volumetric flux on a surface [40], the following expression can be used:

$$\overline{Q}'' = \frac{Q}{\pi L^2/4}, \quad (1)$$

where Q is the volumetric flow rate and L is the surface length.

The variation in the flow rate was ensured by maintaining the overpressure of the phases in the range from 1 to 3 bars at the nozzle inlets. In addition, the change in the parameters of spray irrigation was controlled by changing the position of the nozzle shut-off valve. In the experiments, three different spray irrigation regimes, so-called intermediate and dense spray [3] with either smooth or highly perturbed liquid film, as will be shown below, characterized by different irrigation patterns, were studied. In Figure 2, the measured values of the volumetric flow rate Q and the mean volumetric flux on a surface Q'' for the studied irrigation regimes are shown depending on the liquid overpressure. The minimum studied volumetric flow rate was 0.05 mL/s and the maximum was 1.7 mL/s.

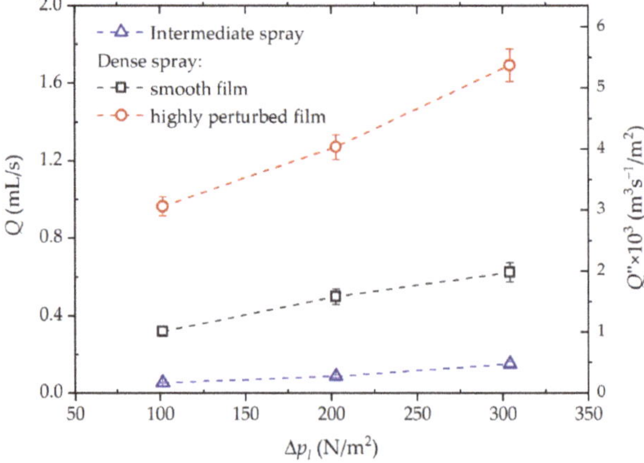

Figure 2. Liquid flow rates for the studied spray cooling regimes. The error bars are presented for the Q values.

As the analysis of the literature shows, the Stokes number (St) is also one of the key parameters of particle-laden flows, among which are spray flows. In particular, two types of situations can be observed for droplets suspended in gas flow: (a) if the St << 1, then the droplet will have ample time to respond to changes in gas flow velocity; (b) vice versa, for the case of St >> 1, the droplet has essentially no time to respond to the gas velocity changes and the its velocity will be little affected by the gas velocity change. The Stokes number is defined as the ratio of the particle momentum response time over a flow system time:

$$St = \frac{\tau_d}{\tau_f}. \quad (2)$$

For considered in the study system the Stokes number has the following form:

$$St = \frac{\rho_d d_d^2}{18\mu_a} \frac{V_f}{d_0}, \quad (3)$$

where ρ_d and d_d are the density and size of the droplets, V_f is the velocity of a spray flow, μ_a is the dynamic viscosity of the gas phase, and d_0 is the orifice diameter of a spray nozzle. The estimations show that, for the spray flow studied in the paper the value of St number (3) varies in the range of 50–1300, which identifies it as inertial flow. Here, the velocity of droplets was estimated using the high-speed video recording from the side of the spray flow.

2.3. High-Speed Video and Infrared Recordings

To study the spray irrigation patterns, a Phantom VEO 410L camera with a maximum frame rate of 100,000 fps and resolution of 30 µm per pixel was used. The analysis of the obtained video shows that the recording format from the reverse side of a heater allows to provide in one experiment an extensive dataset on the hydrodynamic characteristics of spray irrigation. Figure 3 demonstrates the advantages of using a transparent design of the heating surface and the high-speed visualization from its reverse side. As can be seen from the figure, it becomes possible to study the integral characteristics of irrigation; for example, to determine the area of wetted regions for low liquid flow rates, to estimate the characteristic number of liquid droplets falling onto the impact surface, to analyze the uniformity of irrigation, and so on. In addition, it is possible to identify single droplets before their interaction with either the impact surface or the forming liquid film. In turn, this allows to conduct a detailed statistical analysis of their size distribution. Finally, the used recording format allows to analyze the features of the interface evolution during the droplets' impacts on the liquid film forming on the heater.

To study the temperature field of the thin-film ITO heater, an FLIR Titanium HD 570 M high-speed infrared (IR) camera with a frame rate of 1000 fps and a resolution of 120 µm per pixel was used. Before the experiments, the calibration procedure for the IR camera was performed using a small-sized Honeywell temperature sensor located near the heater.

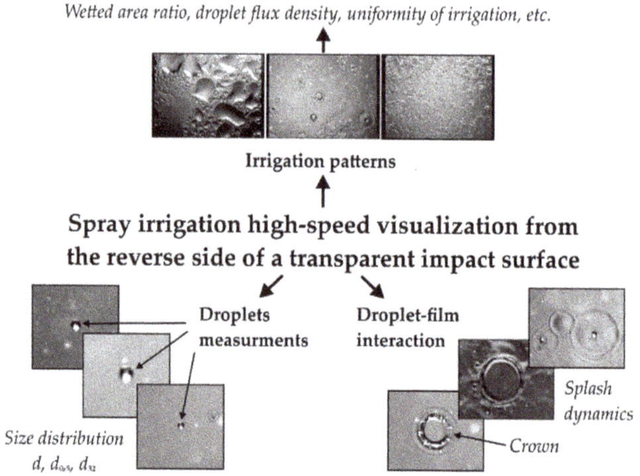

Figure 3. Prospects of using a transparent impact surface for studying the hydrodynamic parameters of spray irrigation.

2.4. Measurement Uncertainties

As the sizes of droplets were analyzed using high-speed video recording data, the uncertainty of such measurements is directly associated with the spatial resolution of the used camera. In turn, the errors of determining both average diameters and number (to analyze droplet flux density) of droplets for a given flow rate directly depend on the statistical error. To minimize it, the wide ensembles of droplets (up to 200–300 droplets) were analyzed in the study. The uncertainty of the time characteristics measurements is associated with the temporal resolution of the video camera, which was at least 0.05 ms. To study the features of the droplet–film interaction, the recording frequency was 100,000 fps, which provided the temporal resolution of 0.01 ms.

The maximum error in determining the value of the volumetric flow rate Q was 5%. The uncertainty of the heat flux density measurement includes an uncertainty associated with the current passed through the ITO heater and the corresponding voltage measurements. In total, their contribution is about 0.3%. However, to determine the real heat flux density supplied to the impact surface, it is necessary to evaluate the heat losses, which make the main contribution to the q value measurements' uncertainty. Based on the calculations made in Comsol Multiphysics, they were estimated to be less than 3% for the used heater design. Finally, according to the mentioned calibration procedure, the absolute error in measuring the surface temperature in the experiments did not exceed 1 °C.

3. Results

3.1. Irrigation Patterns

Using high-speed video recording from the reverse side of the impact surface, the irrigation patterns were studied for various liquid flow rates under adiabatic conditions (Figure 4). From the presented frames, it can be seen that, depending on the flow rate, there is a significantly different nature of the irrigation. In particular, for the so-called intermediate spray, the amount of liquid falling on the surface is not enough to form a continuous liquid film [3]. In this regime, the formation of the separate wetted areas up to 10 mm in size is observed on the impact surface. The coalescences of neighboring liquid areas into larger ones and their subsequent runoff from the surface under the action of gravitational forces are periodically observed. The second regime with a higher liquid flow rate is characterized by the continuous smooth liquid film formation, which is locally disturbed by falling droplets. In the presented frames, it is possible to clearly trace the interface evolution during the droplets' impacts on the film. For the next studied regime, a liquid film with a significantly perturbed interface is observed on the impact surface, which is primarily caused by a significant increase in the droplet flux density.

Intermediate spray

Dense spray with the formation of smooth liquid film

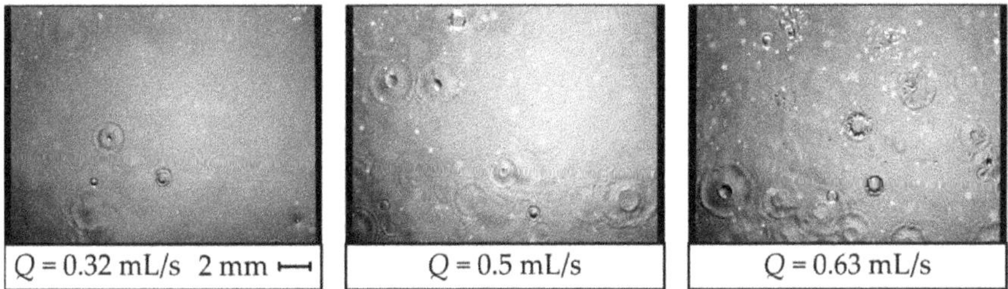

Dense spray with the formation of highly perturbed liquid film

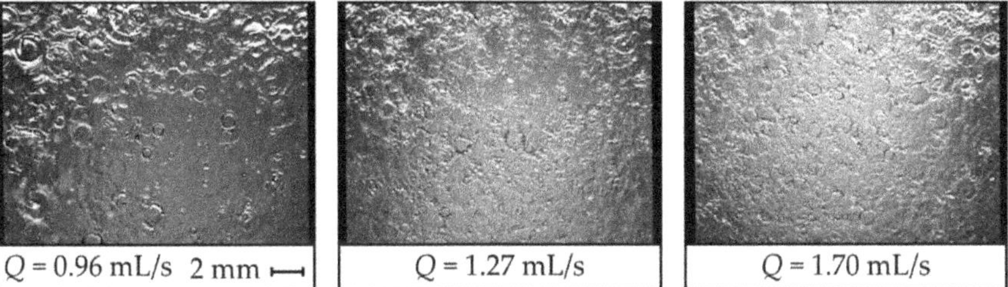

Figure 4. The various patterns during spray irrigation of the transparent impact surface investigated in the paper.

3.2. Droplet Sizes

As shown in Figure 3, owing to the high spatial and temporal resolutions of the used recording format and its focusing features, it is possible to identify and measure the sizes of droplets before their direct impact on the liquid film. This, in turn, makes it possible to obtain a detailed droplets' size distribution for various regimes. As an example, Figure 5a shows a histogram of the droplet diameter distribution for a liquid flow rate Q = 0.5 mL/s. It indicates that droplet sizes can vary over a wide range, from 40 to 170 μm for the presented case.

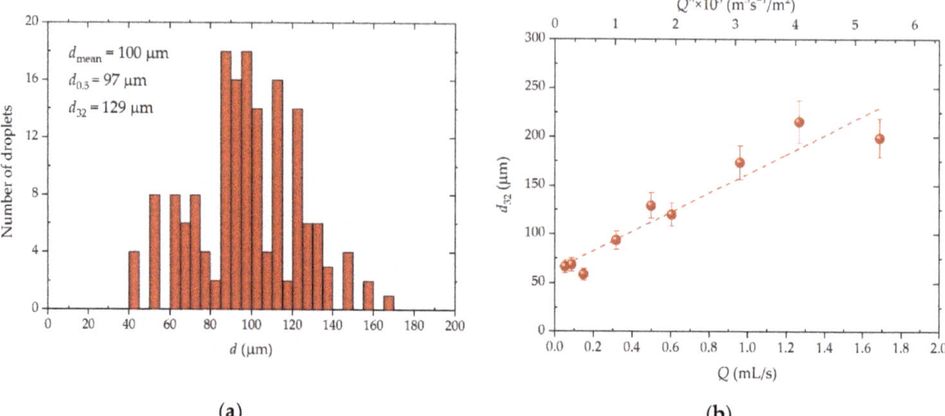

Figure 5. (**a**) Droplet size distribution during spray irrigation (Q = 0.5 mL/s); (**b**) Sauter mean diameter dependence on the flow rate.

Using the obtained dataset, the values of the mass median diameter ($d_{0.5}$) and Sauter mean diameter (d_{32}) were determined, the last of which is defined as follows [41]:

$$d_{32} = \frac{\sum_i n_i d_i^3}{\sum_i n_i d_i^2}, \tag{4}$$

where n_i is the number of droplets with a diameter of d_i. Figure 5b demonstrates the dependence of the d_{32} value on the liquid flow rate. It can be seen that, with an increase in the liquid flow rate in the investigated range, the Sauter mean diameter increases and the obtained dependence d_{32} (Q) with an accuracy of ±25% is described by a linear function.

Certainly, the limit of the droplet diameter measurements in the considered visualization format is directly related to the spatial resolution of the video recording. With the use of special macro lenses, this parameter can be significantly improved; for example, as was done for visualizing individual bubbles during boiling on the surface of a microheater (resolution was 16 μm/pix) [42]. Therefore, in the authors' view, the recording format used is a promising method for measuring small-sized droplets of spray. It can be good alternative in some tasks to modern techniques such as laser diffraction, interferometric particle imaging, and others mentioned in the introduction.

3.3. Droplet Impact on Liquid Film

A particular interest of researchers today is associated with the evolution of the interfacial surface of a liquid film during the impact of individual droplets [43]. This interest is primarily owing the fact that the drop impact on the forming liquid film is one of the key mechanisms of spray cooling and plays an important role in the heat and mass transfer processes [44]. In particular, it can cause a sharp decrease in the impact surface temperature at the area of interaction. In this case, for a certain droplet size and velocity (We number), its impact leads to the formation of a so-called residual film [45]. In turn, this causes the enhanced convective heat transfer and enhanced synergy of velocity field and heat flow field inside the residual film [46]. It is interesting to note that this process is similar in mechanism to the sharp decrease in surface temperature during the nucleation during boiling. In this case, the evaporation of a thin layer of liquid formed at the base of the vapor bubble—the so-called microlayer [47]—leads to high local heat flux densities, exceeding up to 40 times the supplied heat flux [24,28,48].

The transparent design of the impact surface allows to study the scenarios of droplet–film interaction during spray cooling. Indeed, as can be seen from Figure 6, at a given liquid flow rate (Q = 0.5 mL/s), various scenarios of the interface evolution are observed. Most of the small droplets (less than 100 μm) cause capillary wave disturbances on the film (Figure 6a). In this case, it seen that there is no noticeable change in the thickness or structure of the film. Moreover, the oscillations of the interface decay in a fairly short period of time.

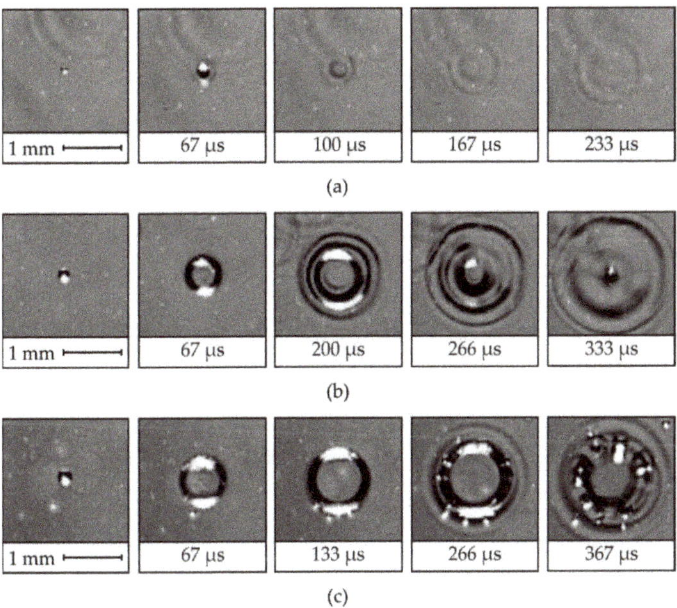

Figure 6. Various scenarios of the droplet impacts on the liquid film observed during spray cooling (Q = 0.5 mL/s): (**a**) small-scale wave disturbance caused by small droplets; (**b**) crater formation; (**c**) splashing crown.

For larger droplets (from 100 μm), the formation of large-scale and relatively long-term perturbations of the film is observed, accompanied by a noticeable change in the interface structure (Figure 6b). In particular, such droplets' impacts lead to the formation of craters with marked ejecta sheet, the depth of which can be comparable to the thickness of the liquid film. The diameter of such craters is up to 2 mm. In addition, large droplets can also cause splashing crowns (Figure 6c), both with and without the secondary droplets [43]. As the analysis of the mentioned literature shows, the last two described scenarios can lead to the formation of residual films. Therefore, droplets leading to such scenarios of the interface disturbance are of greatest interest when considering the mechanisms of heat transfer during spray cooling.

3.4. Droplet Flux Density

The droplet flux density is the number of droplets falling onto a unit of surface area per time unit:

$$\dot{N} = \frac{n}{A\tau}. \tag{5}$$

Similar to nucleation site density for nucleate boiling, the droplet flux density is an extremely important parameter for considering the intensity of mass and heat transfer during spray cooling. At the same time, given the impossibility of accurate measurement of the droplet flux density using the traditional visualization from the side of the impact surface, there are only few studies in the literature devoted to the analysis of this characteristic. This does not allow establishing the relationship of this parameter with the intensity of heat transfer during spray cooling. In turn, the high-speed visualization from the reverse side of a transparent surface makes it possible to investigate this value even for high liquid flow rates. It is also important to note here that it becomes possible to analyze the number of droplets, whose impacts lead to the formation of the residual film described previously.

The determination of the $\dot{N}$ value was performed as follows. For each investigated flow rate, a region with a given area (A = 30 mm^2) was selected, and the number of droplets falling onto the surface within the boundaries of this area was counted for 500–800 ms. It is important to note that this technique makes it possible to count all droplets that come onto the surface, as even small droplets cause wave disturbances of the liquid film, which are clearly distinguishable in high-speed video frames (Figure 6a). The error of determining the droplet flux density in such way for small liquid flow rates directly depends on the statistical error. At high flow rates (more than 1.5 mL/s), counting the number of droplets becomes more difficult owing to the large number of local "events". In this case, the maximum measurement error of the $\dot{N}$ is up to 20%. However, it should be noted here that this value can be reduced, including through the use of automatic image processing algorithms, as well as neural networks, which will increase the sampling and improve the measurement accuracy.

Using the obtained video dataset, the droplet flux density was determined for various liquid flow rates. Figure 7 shows the comparison of the $\dot{N}$ values obtained for the central and edge regions of the surface for the liquid flow rates of Q = 0.32–0.63 mL/s. The figure demonstrates that, with an increase in the flow rate, the described parameter increases linearly. Meanwhile, the droplet flux density can vary significantly depending on the impact surface region selected for the analysis. In particular, for these flow rates, the number of droplets that come to the central region of the impact surface is on average 15% higher than those that come to the edge regions. Obviously, such unevenness is associated with the characteristics of the nozzle used. As will be shown below, this factor can significantly affect the distribution of the temperature field under the conditions of single-phase heat transfer. Thus, in addition to new experimental information about the droplet flux density, the visualization format considered in the study also allows to analyze the uniformity of spray irrigation, which is an extremely important aspect when choosing a nozzle.

3.5. Heat Transfer Rate

The use of high-speed infrared thermography made it possible to analyze the temperature field of the impact surface and to determine the intensity of heat transfer in various modes of spray cooling. For example, in Figure 8, the time-averaged IR thermography frame and the corresponding temperature distribution of the heating surface for a liquid flow rate Q = 1.7 mL/s and heat flux density q = 45 kW/m^2 are presented. The obtained temperature distribution indicates that there is a difference in temperature values at the center and at the periphery of the heating area. However, this difference for the presented case is no more than 3 °C, but it can increase with an increase in the heat flux density. This fact, in our opinion, is primarily owing to the non-uniform distribution of the droplet flux density over the surface, as demonstrated earlier (Figure 7).

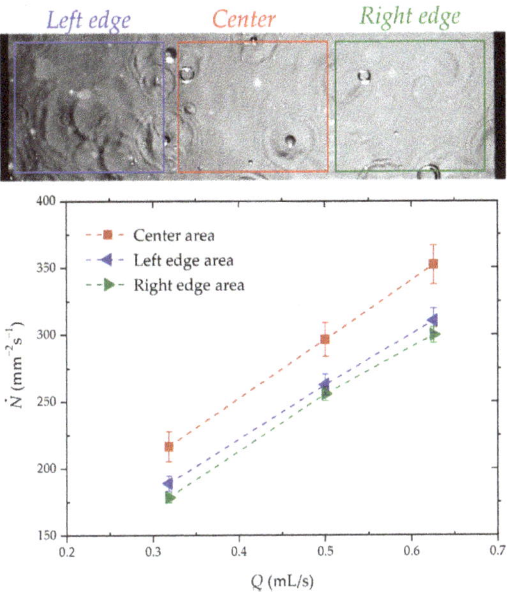

Figure 7. The dependence of the droplet flux density on the liquid flow rate for various impact surface areas.

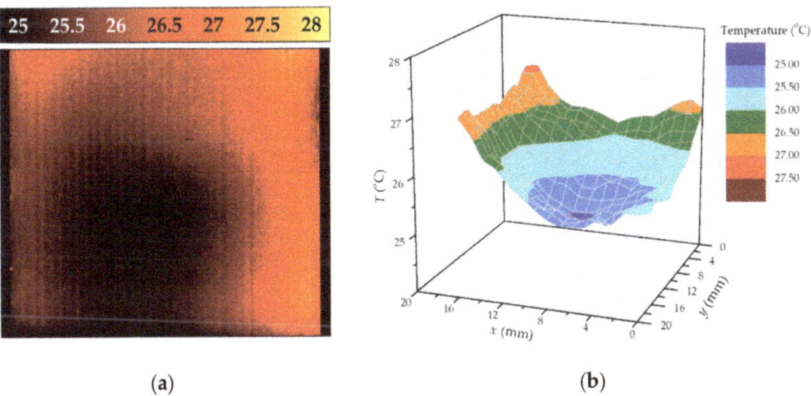

(**a**) (**b**)

Figure 8. (**a**) The time-averaged IR thermography frame and (**b**) corresponding temperature distribution of the heating during spray cooling (Q = 1.7 mL/s, q = 45 kW/m^2).

With the use of the obtained dataset on the surface temperature, the values of the heat transfer coefficient (HTC) for spray cooling were obtained:

$$\mathrm{HTC} = \frac{q}{\overline{T}_s - T_0}, \quad (6)$$

where $\overline{T}_s$ is the surface temperature averaged over time (10 s) and area and T_0 is the initial liquid temperature. Figure 9a shows the dependences of the HTC value on the heat flux density at various liquid flow rates. It seen that the intensity of single-phase heat transfer during spray cooling weakly depends on the heat flux in the studied range. At the same time, an increase in the flow rate leads to a noticeable increase in the intensity of heat transfer (Figure 9b).

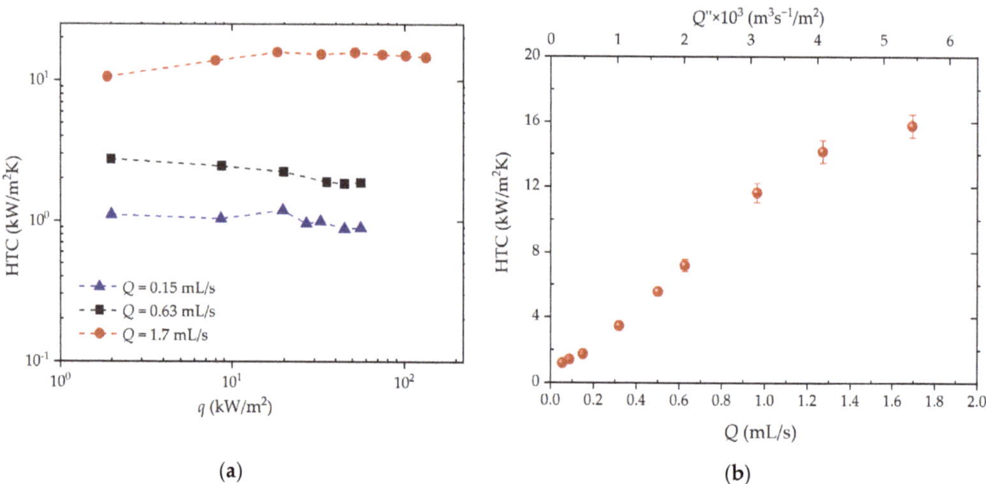

Figure 9. The dependence of the heat transfer coefficient during spray cooling on (**a**) heat flux density and (**b**) liquid flow rate.

At the next stage, the data on the intensity of single-phase heat transfer were compared with the existing correlations (Figure 10). For analysis, the models of Mudawar, Valentine [49], and Rybicki and Mudawar [50] were selected, which have the corresponding forms:

$$\mathrm{Nu} = 2.512\mathrm{Re}_s^{0.76}\mathrm{Pr}_f^{0.56}, \tag{7}$$

$$\mathrm{Nu} = 4.70\mathrm{Re}_s^{0.61}\mathrm{Pr}_f^{0.32}. \tag{8}$$

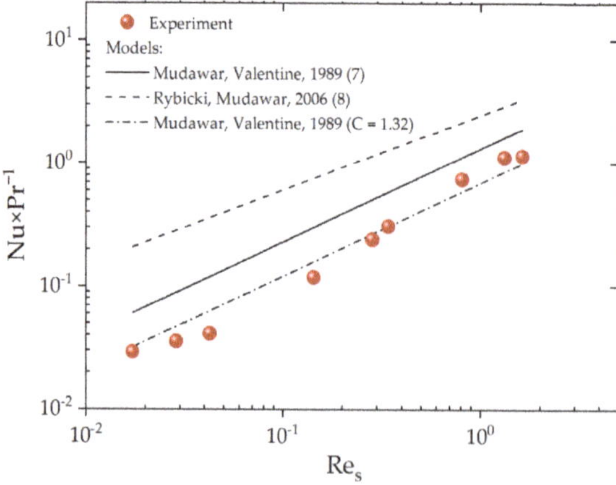

Figure 10. The comparison of the obtained data on the heat transfer intensity during spray cooling with the existing correlations (q = 45 kW/m^2).

In these models, mean volumetric flux on a surface (Q'') and Sauter mean diameter (d_{32}) are used as characteristic scales of velocity and length, respectively:

$$\mathrm{Nu} = \frac{\mathrm{HTC} \cdot d_{32}}{k_f}, \tag{9a}$$

$$\mathrm{Re}_s = \frac{\rho_f Q'' \cdot d_{32}}{\mu_f}, \tag{9b}$$

$$\mathrm{Pr}_f = \frac{C_{pf}\mu_f}{k_f}. \tag{9c}$$

A comparison with the presented models shows that they both demonstrate overestimated values in comparison with the obtained experimental data. Meanwhile, the character of the obtained dependence is close to the correlation [49] and the experimental data can be described within its framework when the numerical constant in expression (7) changes by 1.32.

It is interesting to note that the presented dependences (7) and (8) are close in their form to the criterion dependences used to describe heat transfer during nucleate pool boiling (without taking into account the effect of the properties of heat exchange surface) [51]. In particular, in [52], based on the large dataset generalization, the following dependence was obtained:

$$\mathrm{Nu}_* \sim \mathrm{Re}_*^{0.8}\mathrm{Pr}_f^{1/3}. \tag{10}$$

As can be seen from the comparison, the general view and the exponents at the similar numbers of expression (10) are close to expressions (7) and (8). In addition, as the characteristic dynamic scales of nucleate boiling, the capillary constant and the evaporation rate are used in (10). These parameters are similar to those used for spray cooling description (Figure 9a,b). Moreover, obtained in Figure 9b, the character of the HTC(Q) dependence during spray cooling,

$$\mathrm{HTC} \sim Q^{0.78}, \tag{11}$$

is close to the expressions used to describe heat transfer during nucleate boiling [51]:

$$\mathrm{HTC} \sim q^{0.6-0.8}. \tag{12}$$

Such similarity in the description of these two processes argues in favour of the hypothesis that the physical mechanisms of spray cooling and nucleate boiling are generally similar. In particular, the main contribution to heat transfer during spray cooling, like bubble formation for boiling, is provided by the droplet impact on the liquid film. Some difference in the exponents in the used expressions can be caused by the fact that, during spray irrigation, not all droplets impacting the film have a significant contribution to heat transfer. In particular, as noted earlier, only droplets with a sufficient Weber number form residual film and crown splashing, which lead to a noticeable change in the temperature field of the surface.

4. Conclusions

Using a transparent design of the heating surface and high-speed video and thermographic recordings, a comprehensive study of the hydrodynamic and thermal characteristics of spray cooling was performed with varying liquid flow rate and heat fluxes in a single-phase heat transfer regime. As a result, the following results were obtained:

- The study of the droplet size distribution was carried out and the dependence of the Sauter diameter on the liquid flow rate for the studied irrigation modes was obtained.
- The droplet flux density for various flow rates was studied. It has been shown that this parameter can differ significantly depending on the impact surface region. This

makes it possible to assess the degree of irrigation irregularity even in the case of a full cone spray.

- Various possible scenarios of interaction between droplets and a liquid film forming on the surface were shown, such as the formation of small-scale capillary waves for small droplets, as well as the appearance of craters and splashing crown in the case of large ones.
- Based on the data of high-speed infrared recording, the intensity of heat transfer during spray cooling for various heat flux densities and the liquid flow rates was analyzed. It was shown that, for the studied regimes, the value of the heat transfer coefficient is weakly dependent on the heat flux and is determined primarily by the flow rate.
- Comparison of the obtained experimental data on the intensity of single-phase heat transfer during spray cooling with existing models was performed. It was shown that data can be described within the model of [49] with a modified numerical coefficient. In addition, based on a comparative analysis of existing approaches, an analogy in the mechanisms that determine the intensity of heat transfer during spray cooling and nucleate boiling was shown.

Thus, the performed experimental study shows high prospects of using a transparent design of the heating surface to analyze the main characteristics of spray cooling. In particular, video recording from the reverse side of a transparent impact surface can serve as a good alternative to existing techniques, for example, laser diffraction, interferometric particle imaging, and so on. The authors hope that the information presented in this paper will be useful for further research in this area.

Author Contributions: Conceptualization, A.S. and V.S.; methodology, A.S., V.S. and N.M.; software, N.M.; validation, V.S.; formal analysis, N.M. and V.S.; investigation, V.S. and A.S.; resources, A.S.; data curation, N.M. and V.S.; writing—original draft preparation, V.S.; writing—review and editing, A.S. and V.S.; visualization, N.M. and V.S.; supervision, V.S. and A.S.; project administration, V.S.; funding acquisition, V.S. All authors have read and agreed to the published version of the manuscript.

Funding: The investigation was supported by the Ministry of Science and Higher Education of the Russian Federation (mega-grant No. 075-15-2021-575). The experimental setup was designed and built within the framework of the state contract with IT SB RAS (No. 121031800216-1).

Institutional Review Board Statement: Not applicable.

Informed Consent Statement: Not applicable.

Data Availability Statement: Not applicable.

Conflicts of Interest: The authors declare no conflict of interest.

References

1. Kim, J. Spray cooling heat transfer: The state of the art. *Int. J. Heat Fluid Flow.* **2007**, *28*, 753–767. [CrossRef]
2. Silk, E.A.; Golliher, E.L.; Selvam, R.P. Spray cooling heat transfer: Technology overview and assessment of future challenges for micro-gravity application. *Energy Conv. Mngt.* **2008**, *49*, 453–468. [CrossRef]
3. Liang, G.; Mudawar, I. Review of spray cooling—Part 1: Single-phase and nucleate boiling regimes, and critical heat flux. *Int. J. Heat Mass Transf.* **2017**, *115*, 1174–1205. [CrossRef]
4. Liang, G.; Mudawar, I. Review of spray cooling–Part 2: High temperature boiling regimes and quenching applications. *Int. J. Heat Mass Transf.* **2017**, *115*, 1206–1222. [CrossRef]
5. Mudawar, I. Recent advances in high-flux, two-phase thermal management. *J. Therm. Sci. Eng. Appl.* **2013**, *5*, 021012. [CrossRef]
6. Fabbri, M.; Jiang, S.; Dhir, V.K. A comparative study of cooling of high power density electronics using sprays and microjets. *J. Heat Transf.* **2005**, *127*, 38–48. [CrossRef]
7. Shedd, T.A. Next generation spray cooling: High heat flux management in compact spaces. *Heat Transf. Eng.* **2007**, *28*, 87–92. [CrossRef]
8. Khandekar, S.; Sahu, G.; Muralidhar, K.; Gatapova, E.Y.; Kabov, O.A.; Hu, R.; Luo, X.; Zhao, L. Cooling of high-power LEDs by liquid sprays: Challenges and prospects. *Appl. Therm. Eng.* **2021**, *184*, 115640. [CrossRef]
9. Yang, J.; Chow, L.C.; Pais, M.R. Nucleate Boiling Heat Transfer in Spray Cooling. *J. Heat Transf.* **1996**, *118*, 668–671. [CrossRef]

10. Xu, R.N.; Wang, G.; Jiang, P. Spray Cooling on Enhanced Surfaces: A Review of the Progress and Mechanisms. *J. Electron. Packag.* **2021**, *144*, 010802. [CrossRef]
11. Sijs, R.; Kooij, S.; Holterman, H.J.; Van De Zande, J.; Bonn, D. Drop size measurement techniques for sprays: Comparison of image analysis, phase Doppler particle analysis, and laser diffraction. *AIP Adv.* **2021**, *11*, 015315. [CrossRef]
12. Zhao, X.; Yin, Z.; Zhang, B.; Yang, Z. Experimental investigation of surface temperature non-uniformity in spray cooling. *Int. J. Heat Mass Transf.* **2020**, *146*, 118819. [CrossRef]
13. Salazar, J.P.; Collins, L.R. Two-particle dispersion in isotropic turbulent flows. *Annu. Rev. Fluid Mech.* **2009**, *41*, 405–432. [CrossRef]
14. Balachandar, S.; Eaton, J.K. Turbulent dispersed multiphase flow. *Annu. Rev. Fluid Mech.* **2010**, *42*, 111–133. [CrossRef]
15. Rani, S.L.; Dhariwal, R.; Koch, D.L. A stochastic model for the relative motion of high Stokes number particles in isotropic turbulence. *J. Fluid Mech.* **2014**, *756*, 870–902. [CrossRef]
16. Dhariwal, R.; Rani, S.L.; Koch, D.L. Stochastic theory and direct numerical simulations of the relative motion of high-inertia particle pairs in isotropic turbulence. *J. Fluid Mech.* **2017**, *813*, 205–249. [CrossRef]
17. Dhariwal, R.; Bragg, A.D. Small-scale dynamics of settling, bidisperse particles in turbulence. *J. Fluid Mech.* **2018**, *839*, 594–620. [CrossRef]
18. Ouellette, N.T.; Xu, H.; Bourgoin, M.; Bodenschatz, E. An experimental study of turbulent relative dispersion models. *New J. Phys.* **2006**, *8*, 109. [CrossRef]
19. Ohta, H. Experiments on microgravity boiling heat transfer by using transparent heaters. *Nucl. Eng. Design* **1997**, *175*, 167–180. [CrossRef]
20. Diao, Y.H.; Zhao, Y.H.; Wang, Q.L. Photographic study of bubble dynamics for pool boiling of refrigerant R11. *Heat Mass Transf.* **2007**, *43*, 935–947. [CrossRef]
21. Garrabos, Y.; Lecoutre, C.; Beysens, D.; Nikolayev, V.; Barde, S.; Pont, G.; Zappoli, B. Transparent heater for study of the boiling crisis near the vapor–liquid critical point. *Acta Astronaut.* **2010**, *66*, 760–768. [CrossRef]
22. Kaiho, K.; Okawa, T.; Enoki, K. Measurement of the maximum bubble size distribution in water subcooled flow boiling at low pressure. *Int. J. Heat Mass Transf.* **2017**, *108*, 2365–2380. [CrossRef]
23. Serdyukov, V.; Malakhov, I.; Surtaev, A. High-speed visualization and image processing of sub-atmospheric water boiling on a transparent heater. *J. Vis.* **2020**, *23*, 873–884. [CrossRef]
24. Jung, S.; Kim, H. An experimental method to simultaneously measure the dynamics and heat transfer associated with a single bubble during nucleate boiling on a horizontal surface. *Int. J. Heat Mass Transf.* **2004**, *73*, 365–375. [CrossRef]
25. Surtaev, A.; Serdyukov, V.; Zhou, J. Pavlenko, A.; Tumanov, V. An experimental study of vapor bubbles dynamics at water and ethanol pool boiling at low and high heat fluxes. *Int. J. Heat Mass Transf.* **2018**, *126*, 297–311. [CrossRef]
26. Surtaev, A.; Serdyukov, V.; Malakhov, I. Effect of subatmospheric pressures on heat transfer, vapor bubbles and dry spots evolution during water boiling. *Exp. Therm. Fluid Sci.* **2020**, *112*, 109974. [CrossRef]
27. Gerardi, C.; Buongiorno, J.; Hu, L.W.; McKrell, T. Study of bubble growth in water pool boiling through synchronized, infrared thermometry and high-speed video. *Int. J. Heat Mass Transf.* **2010**, *53*, 4185–4192. [CrossRef]
28. Serdyukov, V.S.; Surtaev, A.S.; Pavlenko, A.N.; Chernyavskiy, A.N. Study on local heat transfer in the vicinity of the contact line under vapor bubbles at pool boiling. *High. Temp.* **2018**, *56*, 546–552. [CrossRef]
29. Ravichandran, M.; Bucci, M. Online, quasi-real-time analysis of high-resolution, infrared, boiling heat transfer investigations using artificial neural networks. *Appl. Therm. Eng.* **2019**, *163*, 114357. [CrossRef]
30. Voulgaropoulos, V.; Aguiar, G.M.; Bucci, M.; Markides, C.N. Simultaneous Laser-and Infrared-Based Measurements of the Life Cycle of a Vapour Bubble During Pool Boiling. In *Advances in Heat Transfer and Thermal Engineering*; Wen, C., Yan, Y., Eds.; Springer: Singapore, 2021; pp. 169–173. [CrossRef]
31. Tanaka, T.; Miyazaki, K.; Yabuki, T. Observation of heat transfer mechanisms in saturated pool boiling of water by high-speed infrared thermometry. *Int. J. Heat Mass Transf.* **2021**, *170*, 121006. [CrossRef]
32. Gao, X.; Kong, L.; Li, R.; Han, J. Heat transfer of single drop impact on a film flow cooling a hot surface. *Int. J. Heat Mass Transf.* **2017**, *108*, 1068–1077. [CrossRef]
33. Gao, X.; Li, R. Impact of a single drop on a flowing liquid film. *Phys. Rev. E* **2015**, *92*, 053005. [CrossRef]
34. Josserand, C.; Thoroddsen, S.T. Drop impact on a solid surface. *Annu. Rev. Fluid Mech.* **2016**, *48*, 365–391. [CrossRef]
35. Kant, P.; Müller-Groeling, H.; Lohse, D. Pattern formation during the impact of a partially frozen binary droplet on a cold surface. *Phys. Rev. Lett.* **2020**, *125*, 184501. [CrossRef] [PubMed]
36. Ersoy, N.E.; Eslamian, M. Phenomenological study and comparison of droplet impact dynamics on a dry surface, thin liquid film, liquid film and shallow pool. *Exp. Therm. Fluid Sci.* **2020**, *112*, 109977. [CrossRef]
37. Yoshida, K.I.; Abe, Y.; Oka, T.; Mori, Y.H.; Nagashima, A. Spray cooling under reduced gravity condition. *J. Heat Transf.* **2001**, *123*, 309–318. [CrossRef]
38. Rini, D.P.; Chen, R.H.; Chow, L.C. Bubble behavior and nucleate boiling heat transfer in saturated FC-72 spray cooling. *J. Heat Transf.* **2002**, *124*, 63–72. [CrossRef]
39. Griffin, A.R.; Vijayakumar, A.; Chen, R.H.; Sundaram, K.B.; Chow, L.C. Development of a transparent heater to measure surface temperature fluctuations under spray cooling conditions. *J. Heat Transf.* **2008**, *130*, 114501. [CrossRef]
40. Mudawar, I.; Estes, K.A. Optimizing and Predicting CHF in Spray Cooling of a Square Surface. *J. Heat Transf.* **1996**, *118*, 672–679. [CrossRef]

41. Mugele, R.A.; Evans, H.D. Droplet size distribution in sprays. *Industr. Eng. Chem.* **1951**, *43*, 1317–1324. [CrossRef]
42. Surtaev, A.; Serdyukov, V.; Malakhov, I.; Safarov, A. Nucleation and bubble evolution in subcooled liquid under pulse heating. *Int. J. Heat Mass Transf.* **2021**, *169*, 120911. [CrossRef]
43. Liang, G.; Mudawar, I. Review of mass and momentum interactions during drop impact on a liquid film. *Int. J. Heat Mass Transf.* **2016**, *101*, 577–599. [CrossRef]
44. Breitenbach, J.; Roisman, I.V.; Tropea, C. From drop impact physics to spray cooling models: A critical review. *Exp. Fluids* **2018**, *59*, 1–21. [CrossRef]
45. Berberović, E.; van Hinsberg, N.P.; Jakirlić, S.; Roisman, I.V.; Tropea, C. Drop impact onto a liquid layer of finite thickness: Dynamics of the cavity evolution. *Phys. Rev. E* **2009**, *79*, 036306. [CrossRef]
46. Liang, G.; Zhang, T.; Chen, L.; Chen, Y.; Shen, S. Single-phase heat transfer of multi-droplet impact on liquid film. *Int. J. Heat Mass Transf.* **2019**, *132*, 288–292. [CrossRef]
47. Cooper, M.G.; Lloyd, A.J.P. The microlayer in nucleate pool boiling. *Int. J. Heat Mass Transf.* **1969**, *12*, 895–913. [CrossRef]
48. Golobic, I.; Petkovsek, J.; Baselj, M.; Papez, A.; Kenning, D.B.R. Experimental determination of transient wall temperature distributions close to growing vapor bubbles. *Heat Mass Transf.* **2009**, *45*, 857–866. [CrossRef]
49. Mudawar, I.; Valentine, W.S. Determination of the local quench curve for spray-cooled metallic surfaces. *J. Heat Treat.* **1989**, *7*, 107–121. [CrossRef]
50. Rybicki, J.R.; Mudawar, I. Single-phase and two-phase cooling characteristics of upward-facing and downward-facing sprays. *Int. J. Heat Mass Transf.* **2006**, *49*, 5–16. [CrossRef]
51. Thome, J.R.; Kim, J. *Encyclopedia of Two-Phase Heat Transfer and Flow II: Special Topics and Applications*; World Scientific Publishing Co. Pte. Ltd.: Singapore, 2015; pp. 62–79. [CrossRef]
52. Gogonin, I.I. The dependence of boiling heat transfer on the properties and geometric parameters of heat-transfer wall. *High Temp.* **2006**, *44*, 913–921. [CrossRef]

 water

Article

Droplet Evaporation in a Gas-Droplet Mist Dilute Turbulent Flow behind a Backward-Facing Step

Maksim A. Pakhomov * and Viktor I. Terekhov

Laboratory of Thermal and Gas Dynamics, Kutateladze Institute of Thermophysics, Siberian Branch of Russian Academy of Sciences, Academician Lavrent'ev Avenue 1, 630090 Novosibirsk, Russia; terekhov@itp.nsc.ru
* Correspondence: pakhomov@ngs.ru

Abstract: The mean and fluctuation flow patterns and heat transfer in a turbulent droplet-laden dilute flow behind a two-dimensional single-side backward-facing step are numerically studied. Numerical simulations are performed for water droplets, with the inlet droplet diameters d_1 = 1–100 μm; they have a mass fraction of M_{L1} = 0–0.1. There is almost no influence of a small number of droplets on the mean gas flow and coefficient of wall friction. A substantial heat transfer augmentation in a droplet-laden mist-separated flow is shown. Heat transfer increases both in the recirculating flow and flow relaxation zones for fine, dispersed droplets, and the largest droplets augment heat transfer after the reattachment point. The largest heat transfer enhancement in a droplet-laden flow is obtained for small particles.

Keywords: gas-droplet turbulent flow; backward-facing step; droplet vaporization and dispersion; Eulerian modeling; heat transfer

Citation: Pakhomov, M.A.; Terekhov, V.I. Droplet Evaporation in a Gas-Droplet Mist Dilute Turbulent Flow behind a Backward-Facing Step. *Water* **2021**, *13*, 2333. https://doi.org/10.3390/w13172333

Academic Editor: Giuseppe Pezzinga

Received: 27 July 2021
Accepted: 23 August 2021
Published: 26 August 2021

Publisher's Note: MDPI stays neutral with regard to jurisdictional claims in published maps and institutional affiliations.

1. Introduction

Gas-droplet confined backward-facing step (BFS) flows are usually inhomogeneous and anisotropic. The multiscale interactions between tiny particles (droplets) and the separated turbulent carrier flow represent a complicated process with numerous insufficiently investigated crucial points [1]. The dispersion of liquid droplets, their heating up and their evaporation can complicate the study of the mean and fluctuation flow structure and the dispersed phase distribution in such flows. Turbulent droplet-laden flows behind a BFS are observed in many applications, such as cyclonic separation, heat transfer, flame stabilization in combustors, pneumatic transport, etc. They represent a typical shear flow with a few zones: the core main flow, shear-layer, recirculation, reattachment and flow relaxation regions [2–5]. Each region has its typical length and time macro- and microscales. Our current idea of the flow structure and heat transfer is far from being completed, even for single-phase flows [1–4]. Particle-laden turbulent BFS isothermal flows were experimentally studied in [6,7]. A particle-laden separated isothermal flow was simulated numerically studied in [8–13]. The Lagrangian [8] and Eulerian [9–13] approaches were used to simulate the dispersed phase. The gas-phase turbulence was predicted by the two-equation isotropic turbulence model [8–11,13] and the algebraic Reynolds stress model [12].

The addition of droplets to a single-phase turbulent flow and their evaporation causes significant heat transfer augmentation [14]. Evaporation of droplets and combustion of various liquids under different conditions were experimentally, theoretically and numerically investigated. The droplet diameter and mass fraction mainly determine the rate of their evaporation. The knowledge of complicated and related processes of droplet evaporation in turbulent recirculating flows allows for a more effective control of the flow structure, wall friction and heat transfer in energy equipment. This is essential for successful design, determining optimal operating conditions and reducing pollutant emissions in various industrial applications. Other methods for controlling the mean flow patterns, turbulence level, pressure drop, friction and heat transfer rate are presented by surface modifications (a corrugated wall surface) [15].

The average Stokes number Stk = τ/τ_f is a non-dimensional parameter defining how the particle or droplet interacts with the mean carrier-phase flow [7,16,17]. Here, $\tau = \rho_L d_1^2 (18\mu W)$ is the particle relaxation time, taking the deviation from the Stokes power law, $W = 1 + \mathrm{Re}_L^{2/3}/6$, where $\mathrm{Re}_L = |\mathbf{U}_S - \mathbf{U}_L| d_1/\nu$ is the Reynolds number based on the dispersed phase diameter and $\tau_f = 5H/U_1$ is the time macroscale (characteristic time of fluid motion) [7,16,17]. ρ_L is the dispersed phase density, d_1 is the droplet diameter at the inlet, $\mathbf{U}_S$ and $\mathbf{U}_L$ are the fluid (gas) velocity seen by the droplet and mean droplet velocity, respectively, and μ and ν are the dynamic and kinematic viscosities of the gas phase, respectively. The tiny particles or droplets (Stk < 1) cause an attenuation of the level of turbulent kinetic energy (TKE) of the gas phase, and they interact well with the motion of the carrier flow [6,7]. The large particles (Stk > 1) cause additional turbulence generation due to the formation of vortices caused by the flow around large particles. The Stokes number $\mathrm{Stk}_K = \tau/\tau_K$, based on the Kolmogorov time scale τ_K, is another important dimensionless parameter for describing the behavior of the dispersed phase in a two-phase flow [7,17,18].

Only a few papers have investigated a non-isothermal droplet-laden backward-facing step flow, as well as performed experimental [16,19] and numerical [16] studies. The streamwise velocities, turbulent kinetic energy of phases, dispersed phase mass flux and heat transfer were measured and predicted in [16]. The measurements were performed using the phase Doppler anemometer for water droplets with a mean initial size d_1 = 60 µm, mass concentration M_{L1} = 0.04 and Reynolds numbers $\mathrm{Re}_H = U_m H/\nu$ = (0.5 and 1.1) × 10^4. The study was performed at two heights of the step, H = 10 and 20 mm, and the expansion ratios were ER = $(H + h_1)/h_1$ = 1.14 and 1.29, respectively. Numerical simulations were performed using the RANS model with a standard $k - \varepsilon$ model. The droplets' motion was simulated using the Lagrangian stochastic approach. A considerable increase in heat transfer (more than twofold in comparison with a single-phase separated flow) was obtained. It was found in Reference [19] that the heat transfer coefficient increases significantly (almost doubling) when the gas-droplet mist is added at Re_H = (1.25 and 2.5) × 10^4. The mean size of droplets at the inlet was d_1 = 10 µm, and the mass fraction was M_{L1} = 0.015. The study was performed for two heights of the step, H = 10 and 40 mm, and in two ducts with h = 10 and 60 mm before expansion. The expansion ratios were ER = $(H + h_1)/h_1$ = 2 and 1.67.

The large-eddy simulation (LES) is one of the modern methods for predicting turbulent flows, and it has been used for the modeling of two-phase separated flows [20]. Lagrangian or Eulerian approaches can be used to predict the dispersed phase. However, the use of LES requires high-performance supercomputers, and this restricts the use of large-eddy simulations for engineering purposes. One of the methods allowing for the partial consideration of the anisotropy of the carrier phase in two-phase backward-facing step flows is the use of second-moment closure (SMC) [21].

The main purpose of this work is the numerical prediction of the mean and turbulent flow structure and heat transfer in a gas-droplet flow behind a backward-facing step. This study is a continuation of previous works [22,23]. The numerical model developed in [22] employed a two-fluid Eulerian model and in-house code to simulate droplet-laden separated flow in a pipe with sudden expansion. The governing equations were written and numerically solved of the axisymmetric droplet-laden mist flow in a pipe with sudden expansion. In this paper, we are carried out the numerical investigation of the flow and heat transfer behind a single-side backward-facing step. This is the main difference between this paper and our previous works [22,23]. The RANS and second-moment closure, taking into account the effect of the dispersed phase addition, were used in the numerical model. The influence of droplet mass fraction, inlet Reynolds numbers and particle diameters at the duct inlet on characteristics of the gas-droplet separated flow were investigated in previous works [22,23].

2. Mathematical Model and Method of Numerical Realization

Turbulent gas-droplet mist flow is simulated using a set of steady-state, incompressible two-dimensional RANS equations. The mean gas flow is described by the continuity, which is two momentum in the streamwise and transverse directions, as well as energy equations and the equation of steam diffusion. The Eulerian approach [24,25] is used for the simulation of the motion and heat transfer of droplets. The sets of mean and fluctuating equations for both phases are described in detail [22]. The elliptic-blending second-moment closure [20] is used for the predictions of the gas-phase turbulence, taking into account the effect of the dispersed phase on the carrier turbulence [26,27]. The droplet-laden flow occurs in a dilute regime ($\Phi_1 < 10^{-4}$), and the droplets are fine ($d_1 < 100$ μm). The effect of inter-particle collisions is neglected in the two-phase flow [18]. The droplet–droplet collisions are ignored in such a case, but the effect of the dispersed phase on the gas turbulent flow cannot be ignored [18,28]. The regime of turbulence modification has been identified for $\Phi = 10^{-6}$–10^{-3}, and it is called "two-way coupling" [28,29]. The r.m.s. velocities and temperature pulsations and the turbulent heat flux of the dispersed phase are simulated using the model of [26,27].

Droplet breakup takes place when $\mathrm{We} = \rho(\mathbf{U}_S - \mathbf{U}_L)^2/\sigma \geq \mathrm{We}_{cr} = 7$ [30], but for all droplet diameters at the inlet studied in the present paper, We << 1. Here $U_S = U + \langle u'_S \rangle$ and $\mathbf{U}_L$ are the fluid (gas) velocity seen by the droplet [12,18] and mean droplet velocity, where U is mean gas velocity (derived from RANS) and $\langle u'_S \rangle$ is the drift velocity between the fluid and the particles [18]. Breakup and droplet deformation were not observed in the turbulent gas-droplet flow. Droplet fragmentation at its contact with a duct wall was not considered. The predictions are carried out for the water droplets and air (gas-phase) flow at the inlet cross-section at a uniform wall temperature ($T_W = 373$ K). The temperature inside the droplet radius stays constant since the Biot number is $\mathrm{Bi} = \alpha_L d_1/\lambda_L << 1$ [22] and the Fourier number is $\mathrm{Fo} = \tau_{eq}/\tau_{evap} << 1$ [31], and a droplet evaporates at the saturation temperature. Here, τ_{eq} is the period of existence of an internal temperature gradient inside a droplet and τ_{evap} is the droplet lifetime (the time until complete droplet evaporation). The boundary condition on the heated wall for the droplets correlates to the so-called "absorbing surface" [32]. According to this condition, droplets do not return to the flow after making contact with the solid wall. The wall surface is always dry, and droplets deposited from the droplet-laden flow momentarily evaporate and do not form a liquid film or spots on the wall [22,23]. This assumption is valid for the heated surface when the temperature difference between the wall and the droplet is large enough ($T_W - T_L \geq 40$ K) [31]. The effect of growth of steam bubbles on the wall surfaces was not taken into account. The same assumptions were used in our previous recent numerical simulations [22,23] for droplet-laden mist flow in the pipe with sudden expansion. It may be is important in other thermal boundary conditions on the wall [33].

3. Numerical Procedures and Validation

The second-order upwind finite control volumes approach on a staggered grid is used for the numerical solution of all of the governing equations for the gas and dispersed phases. The convective terms are discretized using the third-order QUICK algorithm, and the diffusion terms are numerically solved by employing the second-order central difference scheme. The SIMPLEC scheme is used for the coupling of velocity and pressure.

The first computational cell is set at a coordinate $y_+ = yU_*/\nu = 0.3$–0.5, where y is the distance from the wall, and U_* is the friction velocity of the single-phase flow. A minimum of 10 control volumes were located to resolve the large gradients in the near-wall region, subjected to viscosity ($y_+ < 10$). Grid sensitivity was studied to obtain the optimum grid resolution that provides the grid-independent solution. For all numerical investigations, we used a basic grid with 400 × 100 control volumes along the streamwise and transverse directions. The grid convergence was verified for three grid sizes: "coarse" 200 × 50, "basic" 300 × 150 and "fine" 500 × 150 control volumes. The Reynolds stress components are obtained using the method of [34].

A more refined grid is applied in the recirculation region and in the zones of flow detachment and reattachment and in the inlet region of the duct. The coordinate transformation is suitable for such a two-dimensional problem:

$$\Delta\psi_j = a \times \Delta\psi_{j-1},$$

where $\Delta\psi_j$ and $\Delta\psi_{j-1}$ are the current and previous steps of the grid in the axial or radial directions, respectively, and a = 1.08 (longitudinal direction) and a = 1.05 (transverse direction). At least 10 CVs were generated to ensure the resolution of the mean velocity field and turbulence quantities in the viscosity-affected near-wall region ($y_+ < 10$).

At the first stage, the model was tested with measurements of the longitudinal mean and pulsation velocities [35], the wall friction coefficient and the heat transfer distributions along the duct length [36] in a single-phase separated flow behind a backward-facing step. A reasonable agreement was obtained between the computed and measured results (the difference is up to 15%), which served as the basis for simulations of the gas-droplet separated flow.

4. The Numerical Results and Discussion

The simulations are performed for the droplet-laden flow at atmospheric pressure. The duct height before sudden expansion is h_1 = 20 mm; after expansion, h_2 = 40 mm, the step height H = 20 mm and the expansion ratio ER = (h_2/h_1) = 2 (see Figure 1). The mean-mass gas velocity at the inlet is U_{m1} = 10 m/s, and the Reynolds number $\text{Re}_H = HU_{m1}/\nu \approx 1.33 \times 10^4$. We add the droplets to the hydrodynamically fully developed single-phase air flow in the inlet cross-section (the section of sudden expansion) and keep the initial droplet velocity constant across the duct height: $U_{L1} = 0.8U_{m1}$. The initial size of droplets in our studies is d_1 = 1–100 μm, and the mass concentration of droplets is M_{L1} = 0–0.1. The vapor mass fraction at the inlet is M_{V1} = 0.005. The temperature of the air and droplets at the inlet is $T_1 = T_{L1}$ = 293 K, and the wall temperature is T_W = const = 373 K. The mean Stokes number is Stk = τ/τ_f = 0.03–2.9, $\tau_f = 5H/U_1$ = 0.01 s, and the Stokes number is $\text{Stk}_K = \tau/\tau_K$ = 0.2–19.

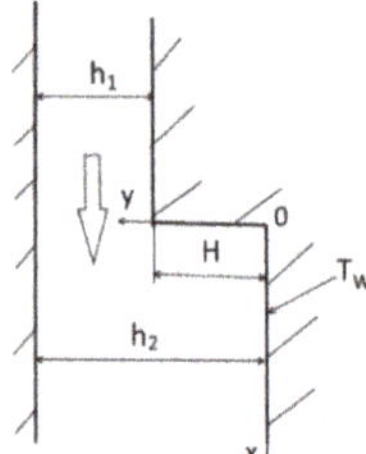

Figure 1. The scheme of the droplet-laden flow in the backward-facing step.

4.1. Flow Structure, Wall Friction and Turbulence Quantities

The transverse profiles of the mean longitudinal velocities of gas and dispersed phases at different distances from the flow separation point are shown in Figure 2a. The flow reattachment point is located at $x_R/H \approx 5.8$ for a single-phase air flow and $x_R/H \approx 5.83$ at M_{L1} = 0.05, where x_R is the length of the recirculating region. The first three cross-sections are located in the recirculation zone, the fourth cross-section corresponds to the flow reattachment area and the last two correspond to the droplet-laden flow relaxation area. A sharp change in the flow structure is observed downstream of the separation cross-section. The profiles of the streamwise velocities of phases in a two-phase flow correspond to those for a single-phase flow. The modification of the mean flow velocity with such a small addition of the dispersed phase is not observed. This qualitatively agrees with other conclusions for both gas-droplet [17,22] and gas-dispersed [6,7,9–12] turbulent separated flows. At a large distance from the point of flow reattachment, the two-phase flow takes the form of a fully developed flow in the duct. In the first cross-sections, the gas velocity is higher than the corresponding value for the dispersed phase; this is explained by the initial conditions for the addition of the droplets to the gas phase and their acceleration in the downward direction. Further, the droplet velocity is almost identical to the gas velocity.

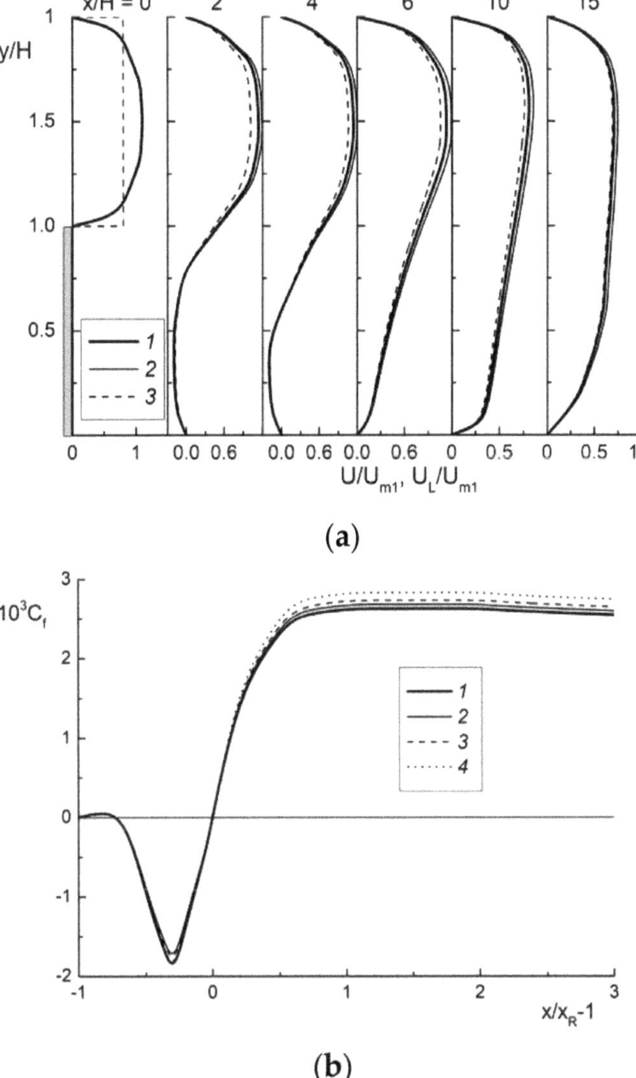

Figure 2. The transverse profiles of (**a**) the mean streamwise phase velocities and distributions of wall friction coefficient along the longitudinal coordinate. d_1 = 30 μm. (**a**): *1* and *2* are single-phase (M_{L1} = 0) and gas-droplet (M_{L1} = 0.05) flows, respectively; *3* is droplets. (**b**): *1*: M_{L1} = 0 (single-phase flow); *2*: 0.02; *3*: 0.05; *4*: 0.1.

The wall friction coefficient ($C_f = 2\tau_W/U_1^2$) distributions along the streamwise coordinate $(x - x_R)/x_R = x/x_R - 1$ are presented in Figure 2b, where τ_W is the shear wall friction and x_R is the recirculation length. The line *1* represents the simulations for the single-phase flow without droplets and with other identical conditions. The addition of a dispersed phase to the turbulent separated single-phase flow has no significant effect on the value of C_f in the flow separation and relaxation (after the reattachment point) regions (see Figure 2b). We can see a slight increase in the wall friction coefficient in a droplet-laden flow.

The distribution of the minimum value of the wall friction coefficient in the separation region for single-phase (2) and gas-droplet flows at M_{L1} = 0.05 (3) vs. the Reynolds number Re_H is given in Figure 3. Line 1 is the semiempirical correlation [37] for calculating the wall friction

$$C_{f,\min} = -0.38 Re_H{}^{-0.57}$$

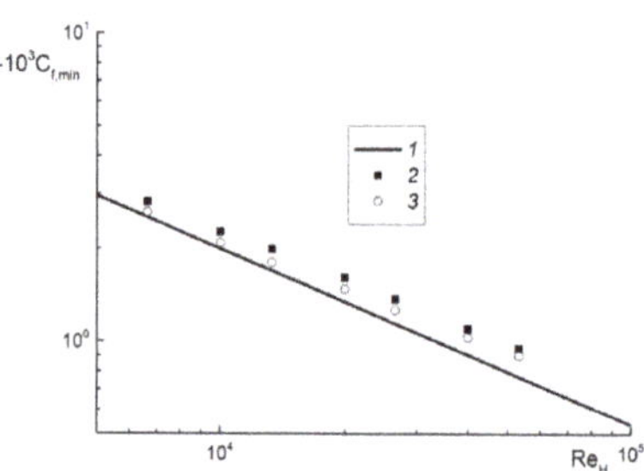

Figure 3. The value of the wall friction coefficient in the recirculation region as a function of the Reynolds number. *1:* semi-empirical correlation of for the single-phase flow [37]; *2:* M_{L1} = 0 (single-phase flow); *3:* 0.05.

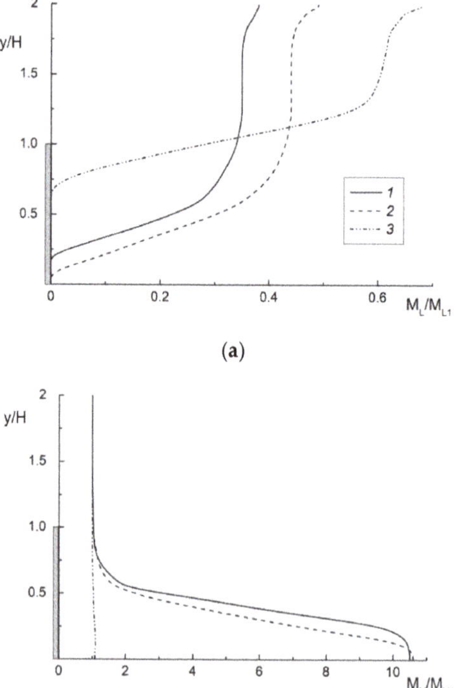

Figure 4. The transverse dictributions of (**a**) droplets and (**b**) steam mass fractions in the droplet-laden separated flow. $x/H = 2$, $M_{L1} = 0.05$. *1:* d_1 = 10 μm, Stk = 0.03; *2:* 30, 0.28; *3:* 100, 2.9.

Points (2 and 3) represent the author's predictions. The effect of droplets on the wall friction in the studied range of the mass fraction is small, up to 5%.

Figure 4a presents the transverse distributions of the droplet mass fraction at $x/H = 2$. The predictions are performed for three different mean Stokes numbers, Stk = 0.03, 0.28 and 2.9. The first two values of mean Stokes numbers (*1* and *2*) correspond to the flow regime with droplet entrainment into the mean turbulent motion, and they are observed in the whole cross-section of the duct (flow core, shear layer and recirculation region). The third Stokes number (3) represents the flow regime, when the droplets are not involved into the mean motion, and they almost do not penetrate into the recirculation zone [6]. These conclusions are confirmed by our numerical simulations. The thin layer close to the duct wall for the small particles is free from droplets due to their evaporation. The height of this zone depends on the droplet diameter, and this height is largest for the smallest particles studied (*1*). The evaporation of droplets in the flow core is insignificant, and the predicted profiles of mass concentrations for all droplet diameters have almost constant values. The slight increase in the profiles of the droplet mass fraction toward the "upper" duct wall is computed. This explains the process of droplet accumulation in the near-wall region and their deposition on the vertical wall surface. The same tendency was numerically obtained in [12] for the gas-dispersed flow behind a BFS.

The droplet mass concentration decreases drastically toward the duct wall. The transverse profiles of vapor mass fractions show the opposite tendency for the initial droplet diameter studied (see Figure 4b). Here, $M_V/M_{V1} = \left(M_{V1} + \Delta M_V^{evap}\right)/M_{V1} = 1 + \Delta M_V^{evap}/M_{V1}$, where ΔM_V^{evap} is the additional mass of steam from the droplet surface created by vaporization. The values of the steam mass concentration are varied from unity (without evaporation) up to 11 at $M_{L1} = 0.05$ (the maximal possible steam magnitude and $M_{L1} \equiv \Delta M_V^{evap}$). The maximum values of the steam mass fraction are found close to the wall of the duct, and the smallest values occur in the duct flow core.

The distributions of the dimensionless diameter of the droplets with a change in their initial size are shown in Figure 5 at a distance $x/H = 2$ downward from the cross-section of the sudden expansion of the flow. A region free of both the smallest (*1*) and the largest (*3*) particles can be seen close to the wall. In the first case, the reason for their absence close to the wall of the duct is the process of their evaporation. For the second case, this is explained by their absence in the entire flow recirculation region due to the large mean Stokes number; such particles are observed only in the shear layer and in the central and "upper" parts of the duct. Droplets of intermediate diameter (2) are observed throughout almost the whole cross-section of the duct. Obviously, in this zone, the droplets' diameters have the smallest values due to the process of their evaporation, whereas in the flow core, their size almost does not differ from the initial value. This confirms the conclusions in Figure 4a.

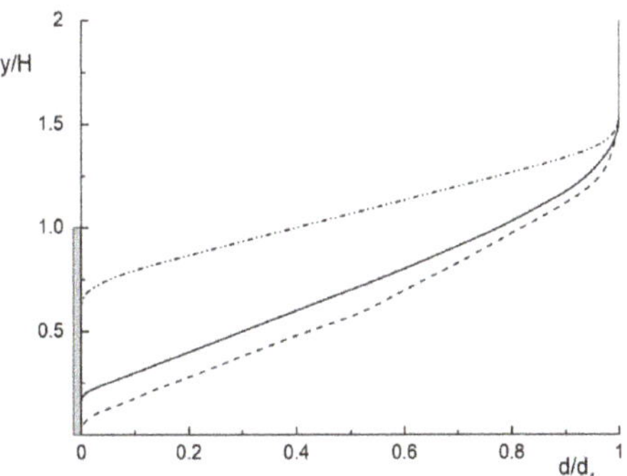

Figure 5. The transverse profiles of normalized droplets diameter in the mist backward-facing step flow. $x/H = 2$, $M_{L1} = 0.05$. Legends to the figure are the same as in Figure 4.

A modification of the gas-phase TKE vs. the mean Stokes numbers at distance $x/H = 2$ is presented in Figure 6. Here, $k_{0,\max}$ is the maximum level of turbulence of the gas phase in a single-phase air flow. The level of gas turbulence in a two-dimensional flow is estimated using the following expression:

$$2k = u'^2 + v'^2 + w'^2 \approx 1.5\left(u'^2 + v'^2\right)$$

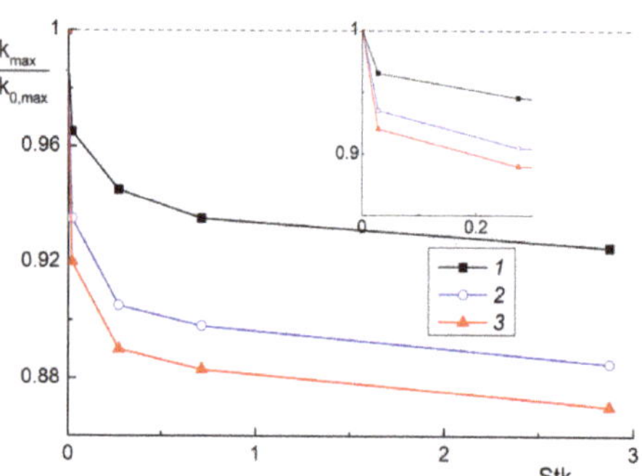

Figure 6. The effect of Stokes numbers on the maximal turbulence kinetic energy modification ratio of the gas-droplet $k_{\max}$ to the single-phase $k_{0,\max}$ phases. $x/H = 2$, d_1 = 0–100 µm. *1:* $M_{L1} = 0$; *2:* 0.05; *3:* 0.1.

The maximum value of the turbulent kinetic energy is predicted in the shear mixing layer in both single- and two-phase flows behind the backward-facing step. The level of turbulence in a two-phase flow attenuates due to addition of tiny droplets in comparison with the single-phase flow. An increase in the droplet mass concentration leads to a significant decrease in the level of the carrier-phase turbulence. This correlates with the previously mentioned experimental [7] and numerical [9,22,23] works for separated flows with solid particles and liquid droplets.

The distributions of the turbulence modification ratio (TMR) k/k_0 depending on the average Stokes number are shown in Figure 7, where k_0 is the turbulence level in the single-phase flow. The carrier-phase turbulence is calculated using relation (11). It is shown that the smallest suppression of gas turbulence is obtained near the wall at the distance $y/H = 0.1$ (*1*), where the diameters of droplets are the smallest due to their evaporation (see Figures 4a and 5). The largest value is predicted at the distance $y/H = 1$ (*3*). There is almost no evaporation in this region, and the diameters of droplets are maximal. They are roughly equal to their initial size. The different mechanisms of the influence of the average Stokes number on turbulence modification for the cross-sections operate in the recirculation zone (*1* and *2*) and after flow reattachment (*3*). There is a sharp bend in the TMR distribution at Stk $\approx$ 1 for lines (*1* and *2*). This is caused by the fact that particles at Stk > 1 scarcely penetrate into the separation region, and they are found only in the shear layer and the duct core. Without a dispersed phase, an increase in turbulence was obtained in these two cross-sections for the level $k/k_0 \to 1$ in a single-phase flow. In the cross-section $y/H = 1$ (*3*), a decrease in the turbulence level of the gas phase is obtained with growth in the mean Stokes number (initial droplet diameter). For the investigated range of the initial droplet diameters, this cross-section is characterized only by turbulence level suppression, while for more inertial particles, an additional generation of the TKE level can also be obtained.

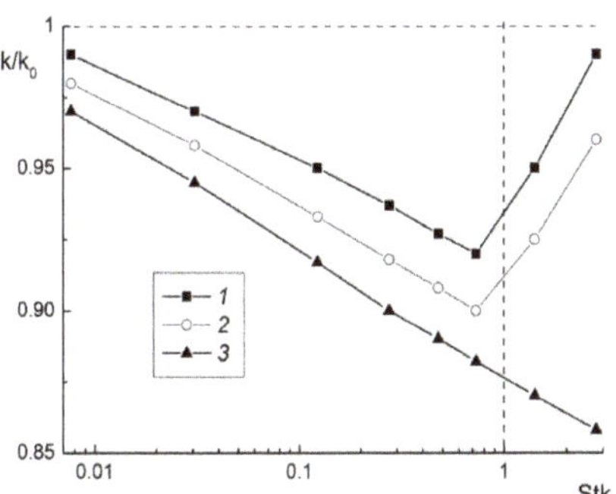

Figure 7. The TKE modification ratio vs. Stokes numbers at a few stations from the wall surface. $x/H = 2$, $M_{L1} = 0.05$. *1:* $y/H = 0.1$; *2:* 0.5; *3:* 1.

The transverse distributions of the temperatures of the gas, $\Theta = (T_W - T)/(T_W - T_m)$ (lines *1* and *2*), and droplets, $\Theta_L = (T_{L,\max} - T_L)/(T_{L,\max} - T_{L,m})$ (line *3*), are presented in Figure 8. Here, T is the temperature and $T_{L,\max}$ is the maximal temperature of the droplets in the corresponding cross-section. The subscripts "*W*", "*m*" and "*L*" correspond to the wall, mean and droplets terms, respectively, and T_m and $T_{L,m} = \frac{2}{U_1 h_2} \int_0^{h_2} T_L U dy$ are the mean-mass gas and droplet temperatures in the corresponding cross-section, respectively. The normalized temperature Θ_L is based on the maximal value of the droplet temperature $T_{L,\max}$. It is obvious that the minimal value of the droplet temperature is predicted in the turbulent flow core, and the maximal value of the droplet temperature is determined close to the wall. The gas temperature in the mist flow (*2*) is lower than that in the single-phase flow (*1*) due to the vaporization of droplets. The droplet temperature in the flow core is slightly lower than at the inlet due to droplet cooling.

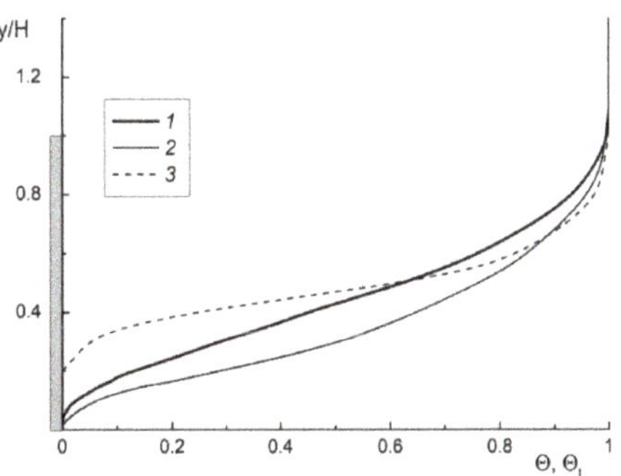

Figure 8. The transverse profiles of gas and droplet Θ_L temperatures in the droplet-laden mist flow in a backward-facing step. d_1 = 30 mm, Stk = 0.28, $x/H = 2$. *1:* single-phase flow ($M_{L1} = 0$); *2:* gas phase ($M_{L1} = 0.05$): *3:* water droplets.

4.2. Heat Transfer

The distributions of local Nusselt numbers along the streamwise coordinate are shown in Figure 9. The Nusselt number at T_W = const is obtained by the following formula:

$$\mathrm{Nu} = \frac{-(\partial T/\partial y)_W H}{T_W - T_m}$$

where $T_m = \frac{2}{U_1 h_2}\int_0^{h_2} TUdy$ is the mean-mass gas temperature. The considerable heat transfer augmentation (more than twofold) in the two-phase mist flow as compared to the single-phase separated flow is obtained. Heat transfer augmentation is predicted in the flow recirculation and relaxation zones as well for Stk ≤ 1. The heat transfer coefficient is maximal in the flow reattachment zone. The coordinate of the maximal value of heat transfer in a gas-droplet flow corresponds approximately to the flow reattachment point. It should be noted that a similar conclusion is also typical of a single-phase flow [3,4]. The heat transfer reduction is similar to that occurring in a single-phase flow due to the growth in thickness of the dynamic and thermal boundary layers at $(x - x_R)/x_R > 1$, where x_R is the length of flow recirculation. The value $\mathrm{Nu}_{fd} = h_2U_{m2}/\nu \approx 38$ (5) is the magnitude of the heat transfer in a fully developed single-phase air flow.

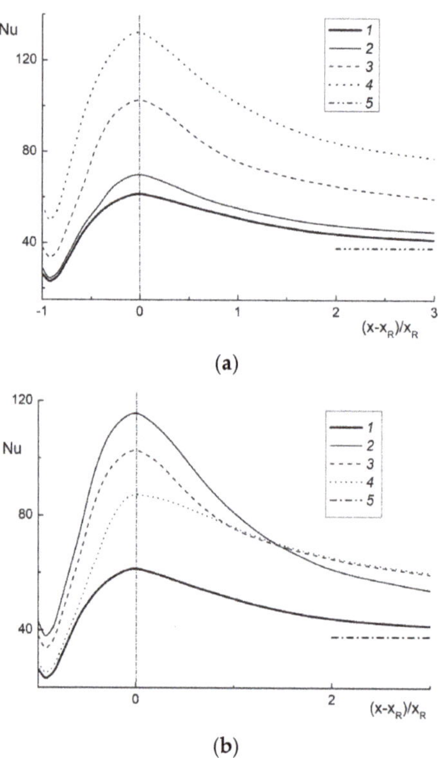

Figure 9. The longitudinal distributions of local Nisselt number for various droplets' inlet mass fraction (**a**) and their diameter. (**a**): d_1 = 30 μm, Stk = 0.28. *1:* $M_{L1} = 0$ (single-phase flow); *2:* 0.02; *3:* 0.05; *4:* 0.1; *5:* the correlation for a fully developed single-phase flow; (**b**): $M_{L1} = 0.05$. *1:* $d_1 = 0$ (single-phase flow); *2:* d_1 = 10 μm, Stk = 0.03; *3:* 30, 0.28; *4:* 100, 2.89; *5:* the correlation for a fully developed single-phase flow.

An increase in the inlet size of droplets has a complicated effect on the heat transfer. The finest droplets (d_1 = 10 μm, Stk = 0.03, 2) evaporate faster and closer to the flow detachment section (see Figure 9b). The latter conclusion confirms the results in Figure 7 showing that droplets are entrained into the separated motion of the gas flow, and they scatter across the recirculating region. The largest droplets at d_1 = 100 μm and Stk = 2.9 (4) are badly entrained into the mean motion of the gas phase. The heat transfer enhancement is revealed mainly after the reattachment point due to droplet evaporation. In the recirculation zone, the values of heat transfer are similar to those in the single-phase separated flow. It should be noted that the maximal value of the Nusselt number in the two-phase flow (d_1 = 100 μm, Stk = 2.9) is smaller than that determined for finer drops (d_1 = 30 μm, Stk = 0.28).

The influence of the average Stokes number on the maximum local value (Figure 10a) and average (Figure 10b) heat transfer is shown in Figure 10. There are two distinctive regions in the Nu_{max} distribution within the whole range of changes in the mass fraction of droplets studied in this work. In the region of small droplet diameters at the inlet (Stk $\approx$ 0.2, $d_1 \approx$ 20 μm), a slight increase in heat transfer is observed, and for large droplets, a sharp reduction is observed. An intensive evaporation of small droplets leads to a decrease in the rate of their deposition. The large droplets almost are not present in the recirculation region. The increase in the droplet mass fraction at the inlet leads to a significant heat transfer enhancement in the two-phase mist flow as compared to a single-phase flow. The largest augmentation of heat transfer in the gas-droplet flow is observed for small droplets.

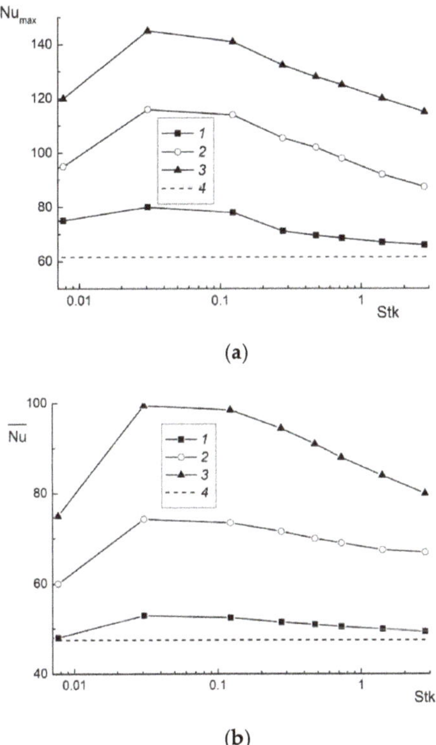

(b)

Figure 10. The effect of mean Stokes numbers (inlet droplets diameter) on (**a**) the maximal value of heat transfer and (**b**) averaged heat transfer. *1:* M_{L1} = 0.02; *2:* 0.05; *3:* 0.1; *4:* single-phase flow.

The influence of the inlet droplet diameter on the average Nusselt number is presented in Figure 10b. The average Nusselt number is calculated as $\overline{\mathrm{Nu}} = \left(\int\limits_0^X \mathrm{Nu}dx\right)/X$, where $X = 25H$ is the length of the averaging zone, and Nu is the local Nusselt number. The magnitude of the average heat transfer for the single-phase flow is $\overline{\mathrm{Nu}_0} \approx 47$. The increase in the droplet mass fraction causes the substantial augmentation of the average Nusselt number (it more than twofold for M_{L1} = 0.1 as compared to the single-phase air flow).

5. Comparison with the Results Obtained for Gas-Dispersed and Droplet-Laden Flows in a Backward-Facing Step

The solid particle-laden turbulent flow behind the backward-facing step was experimentally studied using the laser Doppler anemometer in [7]. The single-side backward-facing step height is H = 26.7 mm, and the duct height before sudden expansion is h_1 = 40 mm. The expansion ratio is ER = 1.67. The Reynolds number based on the maximal centerline velocity U_1 is $\text{Re}_H = HU_1/\nu = 1.84 \times 10^4$. The flow is the single-side backward-facing step oriented vertically downward. The transverse profiles of the gas-phase mean streamwise velocities in particle-laden and single-phase air flows in a few stations were presented in [22], and they are not shown here. Symbols are the experimental results of [7] for glass particles; lines represent simulations of the authors.

Transverse distributions of the ratio of the two-phase to the single-phase longitudinal pulsation velocities of the single-phase air flow are shown in Figure 11 at three cross-sections (x/H = 2, 7 and 14). The predicted suppression of the gas-phase turbulence is noticeable (up to 20% at the coordinate $y/H > 0.5$ for all stations). The gas-phase turbulence modification ratio in the recirculation zone (x/H = 2), in the zone around the reattachment point (x/H = 7) and in the flow relaxation region (x/H = 14) is unchanged for $y/H < 1$. This can be explained by the fact that the mean Stokes number is considerably larger than unity (Stk = 7.3). The magnitude of the particles' mass fraction increases towards the wall due to the effect of turbulent transport (the so-called turbophoresis force) [26,27,32]. The involvement of solid particles in the fluctuation motion and attenuation level of the gas-phase TKE near the duct wall are much smaller as compared to the duct core region.

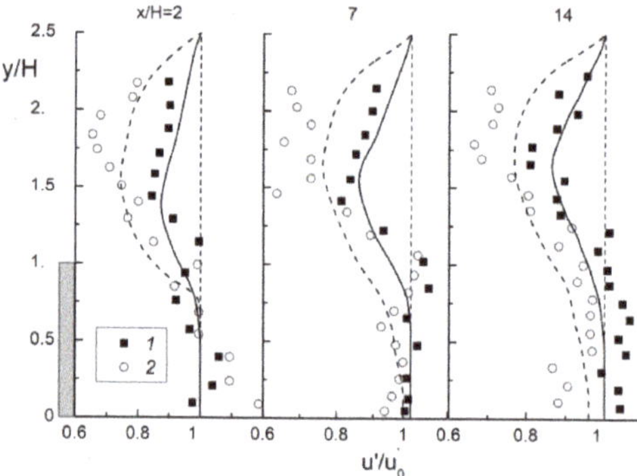

Figure 11. Turbulence modification ratio of the two-phase solid particle-laden flow to the single-phase flow u'/u'_0 longitudinal fluctuating gas-phase velocities. Symbols are the results of experiments of [7] for glass particles, lines are the author's simulations. d_P = 150 μm, ER = 1.67. *1*: M_P = 0.2; *2*: 0.4.

The transverse distributions of the turbulent kinetic energy of the carrier gas and dispersed phases, measured in [16], and the author simulations in several cross-sections behind the BFS are presented in Figure 12. Solid and open symbols are the experimental results of [16]; lines represent author simulations. The turbulent kinetic energy for the gas (air) k and dispersed (water droplets) k_L phases is calculated according to the following approach [16]: $k = 0.5(u'^2 + v'^2)/U_{m1}^2$ and $k_L = 0.5({u_L}'^2 + {v_L}'^2)/U_{m1}^2$. The distributions of turbulence energy for a single-phase flow (*1*) and the gas (*2*) and dispersed (*3*) phases are qualitatively similar. The level of turbulence in a single-phase flow is close to the TKE level in the gas phase of a two-phase gas-dispersed flow. The turbulent energy of droplets is less than the corresponding value for gas (*1* and *2*). There is a maximum in the turbulence level of the gas and dispersed phases, and it is located in the shear layer. This is characteristic of both the experiments of [16] and our numerical results. The largest values of the turbulent kinetic energy for both phases are observed at the distance x/H = 4–5. In the region of the reattachment point ($x_R/H \approx 6.2$), the value of the gas turbulence decreases significantly.

The results predicted by the authors and experiments of [16] are in quantitative agreement, and the largest difference is up to 15%.

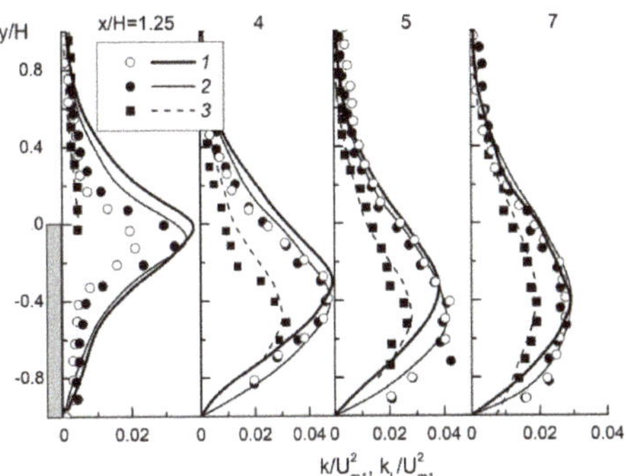

Figure 12. The profiles of turbulent kinetic energies of gas and dispersed phases. Symbols are the results of experiments of [16], lines are the author's simulations. ER = 1.3. *1:* single-phase air flow k_0/U_{m1}^2; *2:* gas phase of droplet-laden mist flow k/U_{m1}^2; *3:* dispersed phase k_L/U_{m1}^2. d_1 = 60 μm, M_{L1} = 0.04, H = 20 mm.

The results of authors' predictions and experiments of [16] concerning distribution of the heat transfer enhancement ratio (HTER) $St/St_{0,max}$ ratios in the mist flow behind a backward-facing step are shown in Figure 13. Here, $St = \alpha/(\rho C_P U_1)$ is the heat transfer rate in a two-phase flow, and $St_{0,max} = \alpha_{0,max}/(\rho_0 C_{P0} U_1)$ is the maximum Stanton number for a single-phase flow behind the BFS, all other parameters being equal. The heat transfer increases by more than 1.5 times in the gas-droplet flow as compared to a single-phase flow, both at the reattachment point and in the flow relaxation zones. Heat transfer enhancement is obtained after the reattachment point at mean Stokes number Stk = 2.2. The value of HTER at Stk = 2.2 (H = 10 mm) in the region of flow relaxation is greater than in the case of a step with Stk = 1.1 (H = 20 mm). The heat transfer enhancement in the recirculation region at Stk = 1.1 is much higher than that at Stk = 2.2. The droplets are well entrained into the recirculation region at a low Stokes number [16]. The local maximum of heat transfer at Stk = 2.2 is significantly below the reattachment point $(x - x_R)/x_R = x/x_R - 1 > 2$, while at Stk = 1.1, the position of the heat transfer maximum almost coincides with the position of the reattachment point of flow.

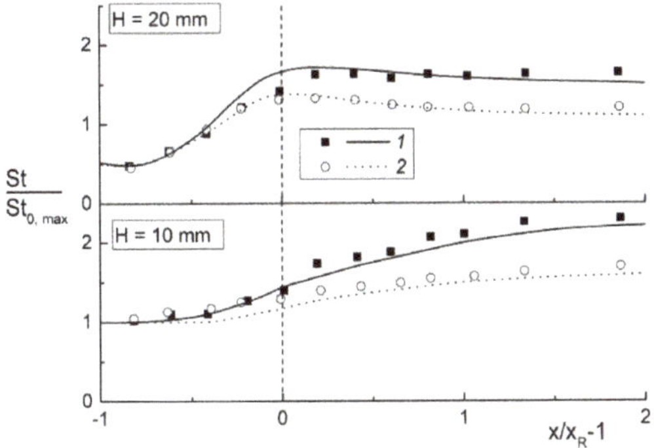

Figure 13. Heat transfer enhancement ratio $St/St_{0,max}$ in the gas-droplets flow behind the backward-facing step for H = 10 and 20 mm. Symbols are the results of experiments of [16], lines are the author's simulations. Um_1 = 10 m/s, $Re_H = Um_1 H/\nu$ = (0.53 and 1.10) × 10^4, d_1 = 60 μm, M_{L1} = 0.04. *1:* T_W = 308 K; *2:* 338 K.

6. Conclusions

The effect of water droplet vaporization on turbulence, flow and heat transfer in the region of the backward-facing step is numerically investigated. The Eulerian two-fluid model is used to simulate the turbulent droplet-laden mist flow. The second-moment closure is modified to take into account the presence of droplets and is predicted to model the turbulent kinetic energy of the carrier gas phase.

The profiles of the longitudinal mean velocities of phases correspond to those for the single-phase flow. The small addition of dispersed phase has almost no effect on the mean flow velocity of the carrier gas flow and the value of the wall friction coefficient. The fine droplets (Stk < 1) are observed in the whole duct cross-section (flow core, shear layer and recirculation region). There are no large droplets with Stk > 1 in the recirculation zone; they are observed in the shear and flow core areas. The suppression of the carrier-phase turbulence (up to 15%) by the addition of fine dispersed droplets is predicted.

Heat transfer in the mist separated flow is significantly increased. The largest increase in heat transfer is predicted for the small droplets. In this case, the heat transfer enhancement is obtained in both the recirculation region and flow relaxation region for the case of fine dispersed particles (Stk < 1). For the large droplets (Stk > 1), heat transfer remains roughly unchanged in the recirculating zone. Heat transfer enhancement is observed only in the reattachment zone.

Author Contributions: Conceptualization, M.A.P. and V.I.T.; methodology, M.A.P. and V.I.T.; Investigation, M.A.P.; data curation, M.A.P. and V.I.T.; formal analysis, M.A.P. and V.I.T.; writing—original draft preparation, M.A.P. and V.I.T.; writing—review and editing, M.A.P. and V.I.T.; resources, M.A.P. and V.I.T.; project administration, V.I.T.; All authors have read and agreed to the published version of the manuscript.

Funding: This work was partially supported by the Ministry of Science and Higher Education of the Russian Federation (mega-grant 075-15-2021-575). The RANS and turbulence model for the single-phase turbulent flow were developed under the state contract with IT SB RAS (121031800217-8).

Institutional Review Board Statement: Not applicable.

Informed Consent Statement: Not applicable.

Conflicts of Interest: The authors declare no conflict of interest.

Nomenclature

$C_f = 2\tau_W/U_1^2$	wall friction coefficient
C_P	heat capacity
d	droplet diameter
ER = $(H + h_1)/h_1$	expansion ratio
h_1	height of the duct before the sudden expansion
h_2	height of the duct after the sudden expansion
H	step height
$2k = \langle u_i u_i \rangle$	turbulent kinetic energy
L	duct length
M_L	mass fraction
$\mathrm{Nu} = -(\partial T/\partial y)_W H/(T_W - T_m)$	Nusslet number
$\mathrm{Re}_H = U_{m1}H/\nu$	the Reynolds number, based on the step height
$\mathrm{Stk} = \tau/\tau_f$	the mean Stokes number
T	temperature
$\mathbf{U}_L$	the mean droplet velocity
U_{m1}	mean-mass flow velocity
$\mathbf{U}_S$	the fluid (gas) velocity seen by the droplet
U_*	friction velocity
$\mathrm{We} = \rho(\mathbf{U}_S - \mathbf{U}_L)^2/\sigma$	the Weber number

x	streamwise coordinate
x_R	position of the flow reattachment point
$x_{\mathrm{Nu\_max}}$	position of the peak of heat transfer rate
y	distance normal from the wall
Subscripts	
0	single-phase fluid (air) flow
1	initial condition
W	wall
L	liquid
m	mean-mass
Greek	
Φ	volume fraction
ε	dissipation of the turbulent kinetic energy
λ	thermal conductivity
ρ	density
μ	the dynamic viscosity
ν	kinematic viscosity
τ	the particle relaxation time
τ_W	wall shear stress
Acronym	
BFS	backward-facing step
CV	control volume
RANS	Reynolds-averaged Navier-Stokes
SMC	second moment closure
TKE	turbulent kinetic energy

References

1. Eaton, J.K.; Johnston, J.P. A review of research on subsonic turbulent flow reattachment. *AIAA J.* **1981**, *19*, 1093–1100. [CrossRef]
2. Simpson, R.L. Aspects of turbulent boundary-layer separation. *Prog. Aerosp. Sci.* **1996**, *32*, 457–521. [CrossRef]
3. Ota, T. A survey of heat transfer in separated and reattached flows. *Appl. Mech. Rev.* **2000**, *53*, 219–235. [CrossRef]
4. Terekhov, V.I.; Bogatko, T.V.; Dyachenko, A.Y.; Smulsky, Y.I.; Yarygina, N.I. *Heat Transfer in Subsonic Separated Flows*; Novosibirsk State Technical University Publishing House: Novosibirsk, Russia, 2016. (In Russian)
5. Chen, L.; Asai, K.; Nonomura, T.; Xi, G.N.; Liu, T.S. A review of backward-facing step (BFS) flow mechanisms, heat transfer and control. *Therm. Sci. Eng. Prog.* **2018**, *6*, 194–216. [CrossRef]
6. Ruck, B.; Makiola, B. Particle dispersion in a single-sided backward-facing step flow. *Int. J. Multiph. Flow* **1988**, *14*, 787–800. [CrossRef]
7. Fessler, J.R.; Eaton, J.K. Turbulence modification by particles in a backward-facing step flow. *J. Fluid Mech.* **1999**, *314*, 97–117. [CrossRef]
8. Chan, C.K.; Zhang, H.Q.; Lau, K.S. Numerical simulation of gas-particle flows behind a backward-facing step using an improved stochastic separated flow model. *J. Comp. Mech.* **2001**, *27*, 412–417. [CrossRef]
9. Zaichik, L.I.; Kozelev, M.V.; Pershukov, V.A. Prediction of turbulent gas-dispersed channel flow with recirculation zones. *Fluid Dyn.* **1994**, *29*, 65–75. [CrossRef]
10. Mohanarangam, K.; Tu, J.Y. Two-fluid model for particle-turbulence interaction in a backward-facing step. *AIChE J.* **2007**, *53*, 2254–2264. [CrossRef]
11. Benavides, A.; Van Vachem, B. Eulerian–Eulerian prediction of dilute turbulent gas-particle flow in a backward-facing step. *Int. J. Heat Fluid Flow* **2009**, *30*, 452–461. [CrossRef]
12. Mukin, R.V.; Zaichik, L.I. Nonlinear algebraic Reynolds stress model for two-phase turbulent flows laden with small heavy particles. *Int. J. Heat Fluid Flow* **2012**, *33*, 81–91. [CrossRef]
13. Riella, M.; Kahraman, R.; Tabor, G.R. Reynolds-averaged two-fluid model prediction of moderately dilute fluid-particle flow over a backward-facing step. *Int. J. Multiph. Flow* **2018**, *106*, 95–108. [CrossRef]
14. Sazhin, S.S. Modelling of fuel droplet heating and evaporation: Recent results and unsolved problems. *Fuel* **2017**, *196*, 69–101. [CrossRef]
15. Abu Talib, A.R.; Hilo, A.K. Fluid flow and heat transfer over corrugated backward facing step channel. *Case Stud. Therm. Eng.* **2021**, *24*, 100862. [CrossRef]
16. Hishida, K.; Nagayasu, T.; Maeda, M. Augmentation of convective heat transfer by an effective utilization of droplet inertia. *Int. J. Heat Mass Transfer* **1995**, *38*, 1773–1785. [CrossRef]
17. Elgobashi, S. On predicting particle-laden turbulent flows. *Appl. Scient. Res.* **1994**, *52*, 309–329. [CrossRef]
18. Zaichik, L.I.; Alipchenkov, V.M. A statistical model for predicting the fluid displaced/added mass and displaced heat capacity effects on transport and heat transfer of arbitrary density particles in turbulent flows. *Int. J. Heat Mass Transf.* **2011**, *54*, 4247–4265. [CrossRef]

19. Miyafuji, Y.; Senaha, I.; Oyakawa, K.; Hiwada, M. Enhancement of Heat Transfer at Downstream of a Backward-Facing Step by Mist Flow. In Proceedings of the 2nd International Conference on Jets, Wakes and Separated Flows ICJWSF-2008, Berlin, Germany, 16–18 September 2008.
20. Wang, B.; Zhang, H.Q.; Wang, X.L. Large eddy simulation of particle response to turbulence along its trajectory in a backward-facing step turbulent flow. *Int. J. Heat Mass Transf.* **2006**, *49*, 415–420. [CrossRef]
21. Fadai-Ghotbi, A.; Manceau, R.; Boree, J. Revisiting URANS computations of the backward-facing step flow using second moment closures. Influence of the numerics. *Flow Turbul. Combust.* **2008**, *81*, 395–410. [CrossRef]
22. Pakhomov, M.A.; Terekhov, V.I. Second moment closure modelling of flow, turbulence and heat transfer in droplet-laden mist flow in a vertical pipe with sudden expansion. *Int. J. Heat Mass Transf.* **2013**, *66*, 210–222. [CrossRef]
23. Pakhomov, M.A.; Terekhov, V.I. The effect of droplets evaporation on turbulence modification and heat transfer enhancement in a two-phase mist flow downstream of a pipe sudden expansion. *Flow Turbul. Combust.* **2017**, *98*, 341–354. [CrossRef]
24. Drew, D.A. Mathematical modeling of two-phase flow. *Ann. Rev. Fluid Mech.* **1983**, *15*, 261–291. [CrossRef]
25. Reeks, M.W. On a kinetic equation for the transport of particles in turbulent flows. *Phys. Fluids A* **1991**, *3*, 446–456. [CrossRef]
26. Derevich, I.V.; Zaichik, L.I. Particle deposition from a turbulent flow. *Fluid Dyn.* **1988**, *23*, 722–729. [CrossRef]
27. Zaichik, L.I. A statistical model of particle transport and heat transfer in turbulent shear flows. *Phys. Fluids* **1999**, *11*, 1521–1534. [CrossRef]
28. Elgobashi, S. An Updated Classification Map of Particle-Laden Turbulent Flows. In Proceedings of the IUTAM Symposium on Computational Approaches to Multiphase Flow, Fluid Mechanics and Its Applications, Argonne National Laboratory, Lemont, IL, USA, 4–7 October 2004; Volume 81, pp. 3–10.
29. Gore, R.A.; Crowe, C.T. The effect of particle size on modulating turbulent intensity. *Int. J. Multiph. Flow* **1989**, *15*, 279–285. [CrossRef]
30. Lin, S.P.; Reitz, R.D. Drop and spray formation from a liquid jet. *Ann. Rev. Fluid Mech.* **1998**, *30*, 85–105. [CrossRef]
31. Snegirev, A.Y. Transient temperature gradient in a single-component vaporizing droplet. *Int. J. Heat Mass Transf.* **2013**, *65*, 80–94. [CrossRef]
32. Derevich, I.V. Statistical modelling of mass transfer in turbulent two-phase dispersed flows. 1. Model development. *Int. J. Heat Mass Transf.* **2000**, *43*, 3709–3723. [CrossRef]
33. Wang, H.G.; Zhang, C.G.; Xiong, H.B. Growth and collapse dynamics of a vapor bubble near or at a wall. *Water* **2021**, *13*, 12. [CrossRef]
34. Hanjalic, K.; Jakirlic, S. Contribution towards the second-moment closure modelling of separating turbulent flows. *Comput. Fluids* **1998**, *27*, 137–156. [CrossRef]
35. Kasagi, N.; Matsunaga, A. Three-dimensional particle-tracking velocimetry measurement of turbulence statistics and energy budget in a backward-facing step flow. *Int. J. Heat Fluid Flow* **1995**, *16*, 477–485. [CrossRef]
36. Vogel, J.C.; Eaton, J.K. Combined heat transfer and fluid dynamics measurements downstream of a backward facing step. *ASME J. Heat Transf.* **1985**, *107*, 922–929. [CrossRef]
37. Tihon, J.; Legrand, J.; Legentilhomme, P. Near-wall investigation of backward-facing step flows. *Exp. Fluids* **2001**, *31*, 484–493. [CrossRef]

Article

The Effect of a Backward-Facing Step on Flow and Heat Transfer in a Polydispersed Upward Bubbly Duct Flow

Tatiana V. Bogatko [1], Aleksandr V. Chinak [2], Ilia A. Evdokimenko [2], Dmitriy V. Kulikov [2], Pavel D. Lobanov [2] and Maksim A. Pakhomov [1,*]

[1] Laboratory of Thermal and Gas Dynamics, Kutateladze Institute of Thermophysics, Siberian Branch of Russian Academy of Sciences, Academician Lavrent'ev Avenue 1, 630090 Novosibirsk, Russia; bogatko1@mail.ru

[2] Laboratory of Problems of Heat and Mass Transfer, Kutateladze Institute of Thermophysics, Siberian Branch of Russian Academy of Sciences, Academician Lavrent'ev Avenue 1, 630090 Novosibirsk, Russia; chinak@itp.nsc.ru (A.V.C.); evdokimenko96@ngs.ru (I.A.E.); kulikov@itp.nsc.ru (D.V.K.); lobanov@itp.nsc.ru (P.D.L.)

* Correspondence: pakhomov@ngs.ru

Abstract: The experimental and numerical results on the flow structure and heat transfer in a bubbly polydispersed upward duct flow in a backward-facing step are presented. Measurements of the carrier fluid phase velocity and gas bubbles motion are carried out using the PIV/PLIF system. The set of RANS equations is used for modeling the two-phase bubbly flow. Turbulence of the carrier fluid phase is predicted using the Reynolds stress model. The effect of bubble addition on the mean and turbulent flow structure is taken into account. The motion and heat transfer in a dispersed phase is modeled using the Eulerian approach taking into account bubble break-up and coalescence. The method of delta-functions is employed for simulation of distributions of polydispersed gas bubbles. Small bubbles are presented over the entire duct cross-section and the larger bubbles mainly observed in the shear mixing layer and flow core. The recirculation length in the two-phase bubbly flow is up to two times shorter than in the single-phase flow. The position of the heat transfer maximum is located after the reattachment point. The effect of the gas volumetric flow rate ratios on the flow patterns and maximal value of heat transfer in the two-phase flow is studied numerically. The addition of air bubbles results in a significant increase in heat transfer (up to 75%).

Keywords: turbulent bubbly flow; backward-facing step; PIV/PLIF measurements; RANS modeling; flow structure; heat transfer

Citation: Bogatko, T.V.; Chinak, A.V.; Evdokimenko, I.A.; Kulikov, D.V.; Lobanov, P.D.; Pakhomov, M.A. The Effect of a Backward-Facing Step on Flow and Heat Transfer in a Polydispersed Upward Bubbly Duct Flow. *Water* **2021**, *13*, 2318. https://doi.org/10.3390/w13172318

Academic Editor: Majid Mohammadian

Received: 29 June 2021
Accepted: 20 August 2021
Published: 24 August 2021

Publisher's Note: MDPI stays neutral with regard to jurisdictional claims in published maps and institutional affiliations.

1. Introduction

Upward gas-liquid bubbly flows in ducts are frequently employed in chemical, nuclear, power engineering and other practical applications. These flows are characterized by a strong interfacial momentum, heat and mass transfer between the carrier fluid phase (water) and dispersed phase (gas bubbles). Bubbly flows in a backward-facing step (BFS) are complicated detachment, recirculation and reattachment of a flow, polydispersity, break-up, coalescence and heat transfer [1,2]. The first papers concerned with the study of such flows were experimental studies of two-phase bubbly flows in ducts and pipes behind a backward-facing step [3,4]. Authors measured the pressure drops, bubble diameter and local void fraction and mean and fluctuational phase velocities.

The complexity of the numerical modeling of such flows is associated with the need to take into account the number of multiscale factors: turbulence of the carrier phase, gas bubble-carrier fluid interaction, bubble break-up and coalescence processes. Only a few papers devoted to the experimental investigations of turbulent bubbly flows in a duct or pipe with sudden expansion without heat transfer were found in the literature [5–7].

A theoretical and experimental investigation of the upward bubbly flow with a sudden pipe expansion was carried out in [5]. The distributions along the pipe length and across the pipe radius of mean and fluctuating axial velocities of the gas bubbles, local void fraction, distribution of bubble diameter, pressure drop and wall friction coefficient were studied in this work. The experimental study of a mixture of air and oil in a horizontal BFS flow was performed in a bubbly flow mode, in a transitional regime from bubbly to annular flows and in an annular regime [6]. It is shown that an increase in the gas volumetric flow rate ratio after the two-phase flow detachment cross-section occurs due to the separation of air from the two-phase flow in the recirculation region.

The flow structure of oil-water and kerosene-water flows in a horizontal conduit behind a sudden contraction and expansion has been experimentally studied [7]. Thick core flow can be revealed for high viscous oil in a sudden expansion. It increases the probability of wall fouling by the oil. The experimental values of contraction and expansion coefficients are found to be lower for oil-water as compared to only single-phase fluid water flow for other conditions being identical. Experimental studies of the pressure drop for a two-phase slug flow in a horizontal channel with a sudden expansion were carried out in work [8]. An electrochemical method was used to measure the wall shear stress. It has been shown that misalignment can affect the shear stress of the wall. It was shown that the gas superficial velocity directly affects the flow behavior, as well as the maximum pressure drop and it standard deviation upstream the sudden expansion of the channel. The experimental investigation using the high-speed photography of gas-liquid flow in a horizontal pipe with a sudden expansion was carried out in [9]. The bubble burst phenomenon is described, and liquid film thickness is estimated.

The PIV and shadowgraph measurements of flow and bubble dynamics in an upward bubbly flow in a square pipe with a sudden expansion were studied in [10]. The experiments were carried out for both laminar and turbulent flow regimes. Authors observed that in a two-phase bubbly flow, smaller bubbles migrate toward the pipe wall and larger ones rise in a tube core region by the effect of steeper velocity gradient with the increasing Reynolds number. The accurate modeling of flow structure, void fraction distribution over a duct cross-section and heat transfer in turbulent bubbly flows with a sudden expansion is very important for safety operations, and the predictions of different emergency situations for the energy equipment of thermal and nuclear power stations.

There are only a few works concerned the numerical study of turbulent bubbly flows behind a pipe or duct with sudden expansion without [11] and with heat transfer [12,13]. The mathematical model for describing a polydispersed bubbly flow in the vertical pipe with sudden expansion and its validation by their own measurements was performed in [8]. The turbulence of the carrier phase was modeled using the k-ω SST model [14]. The bubble break-up and coalescence dynamics are predicted using the inhomogeneous multiple size group (h-MUSIG) model [15].

The mathematical model developed and numerical results of the structure and heat transfer in a polydispersed bubbly turbulent flow downstream of a sudden pipe expansion were presented [12]. The turbulence of the carrier liquid phase is predicted using the model of Reynolds stress transport. Bubble dynamics was predicted taking into account changes in the average volume of bubbles due to the expansion of the change in their density, break-up and coalescence. The study was carried out at the change of initial diameter of air bubbles in the range of d_1 = 1–3 mm and their volumetric gas flow rate ratios of β = 0%–10%. The effect of the gas volumetric flow rate ratios on the flow structure and heat transfer in the two-phase flow is numerically studied in a bubbly polydispersed flow in a horizontal duct with a single-side BFS [13]. The model based on the Eulerian two-fluid approach.

The modeling of vertical turbulent bubbly flows in a pipe (duct) or in a bubble column requires solving a wide range of time and length scales. The large Eddy simulation (LES) is one of the modern methods for simulating turbulent flows without modeling the fluid phase turbulence. This approach for modeling bubbly flows is much more reasonable than DNS, but it is still too complicated for engineering simulations of turbulent bubbly flows. LES was successfully used for the simulation of two-phase bubbly shear flows [16] or bubble motion in a bubble column [17,18] in the last two decades.

The effect of the bubble fraction and their diameter on the length of the flow recirculating zone, heat transfer and carrier fluid turbulence remain open. The authors did not find works concerning experimental or numerical study of heat transfer in turbulent bubbly flows in BFS. This work is a continuation of our previous papers [12,13] for modeling of flow and heat transfer in the bubbly polydispersed turbulent flows in a pipe [12] or duct [13] with sudden expansion. The main difference of this work from [12] is the method of simulations of the evolution of the spectrum of the bubble diameter. In this paper, the method of δ-functions is used [19,20]. Authors [12] used the model of [21] to describe the evolution of bubble size.

Another method for controlling the mean flow patterns, pressure drop, friction and heat transfer rate is the addition of nanoparticles to a backward-facing step flow [22]. The single-phase laminar flow and heat transfer of nanofluid flow in a horizontal backward facing step subjected to a bleeding condition (suction/blowing) is numerically investigated in [22].

The aim of the present paper is the experimental and numerical study of the flow and bubble dynamics, turbulence modification of the carrier fluid phase, mixing and heat transfer enhancement in the upward bubbly flow in a single-side backward-facing step. This paper may be interesting for scientists and research engineers dealing with the problem of flow control and heat transfer enhancement in power equipment.

2. Measurement Setup

The scheme of the test facility is shown in Figure 1a. The test liquid was stored in the tank (1). A centrifugal pump manufactured by Grundfos (Bjerringbro, Denmark) (2) with a maximum flow rate of 4.2 l/s is used to supply the working fluid. The pump speed and the flow rate of the fluid are set by the PumpMaster PM-P540 frequency converter. To measure the flow rate of the test fluid, an ultrasonic flow meter (3) KARAT-520-32-0 (NPO KARAT, Yekaterinburg, Russia) is used, the measurement uncertainty of the flow rate is 0.2% of the measured value. The test section (duct with sudden expansion) (4) was made of an organic glass with the inner dimensions 1000 × 200 × 20 mm. To create a separation zone, a plexiglass plate is fixed inside the channel. The length and height of the duct before the sudden expansion were L1 = 600 mm and h_1 = 8 mm, and after the sudden expansion were L2 = 400 mm and h_2 = 20 mm. The step height H = 12 mm, and the expansion ratio was ER = $(H + h_1)/h_1 = h_2/h_1 = 2.5$. A honeycomb was mounted on the duct inlet in front of the plate to establish a uniform fluid flow. The test section is fixed on a frame made of machine-tool aluminum profiles, which provides convenient fastening of the stand elements and the optical system and easy readjustment.

The liquid flow rate through the working section was 0.72 L/s, which corresponds to the mean-mass flow velocity U_{m1} = 0.55 m/s and Reynolds number based on step height $\mathrm{Re}_H = U_{m1}H/\nu = 6600$. Gas bubbles were introduced into the liquid flow through 9 capillaries with an inner diameter of 0.7 mm located in the lower part of the working section.

The measurements were carried out using the PIV/PLIF system (5), based on the use of a green laser with a constant glow (wavelength 532 nm), full power 1 W and a high-speed camera JET 19 (Kaya Instruments, Haifa, Israel). Since the duct has a large width, measurements in a two-phase flow using the PIV/PLIF method in a situation where the laser knife and the camera's field of view are located perpendicularly is extremely difficult. This is due to the significant overlap of the measurement area by bubbles that move between the region of interest (ROI) and the camera. Therefore, for measurements, the laser knife was wound perpendicular to the wall of the test section, and the camera was directed at an angle of 45 (see Figure 1a). Because of this, we had different scales along the horizontal and vertical axes. The calibration of the linear dimensions was performed for both axes. Measurements were performed along the central axis of the channel with a sudden expansion just behind the BFS (see Figure 1b). The shooting speed during the experiments was 1000 frames per second. Distilled water with the addition of fluorescent (rhodamine filled) polyamide particles manufactured by Dantec Dynamics (Skovlunde, Denmark) (their diameter was 1–20 μm) was used in measurements. To make bubbles invisible, no Rhodamine was added to the test liquid accorded [23].

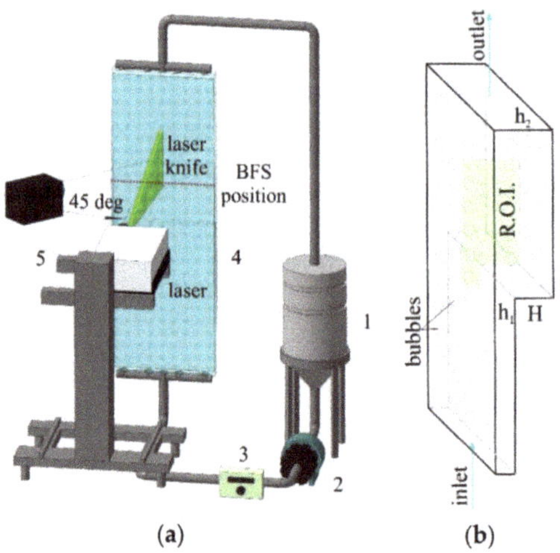

Figure 1. The scheme of the experimental setup (**a**): 1 is the water tank, 2 is the pump, 3 is the flowmeter, 4 is the test section, 5 is the PIV/PLIF system. (**b**) The scheme of a two-phase bubbly upward flow behind a backward-facing step.

The recognition of bubbles in a separated flow is a nontrivial task and requires the development of new methods and approaches. Examples of images obtained by adding Rhodamine to the flow are shown in Figure 2a–c, obtained at Re = 6600 and b = 0.03 and different sizes of bubbles. The figures show that the bubbles in the separation region of the flow can have a substantially non-spherical shape. The currently developed methods, as a rule, deal with spherical or elliptical bubbles [24]. Development of a deep learning-based image processing technique for bubble pattern recognition and shape reconstruction in dense bubbly flows were performed in [25,26]. In addition, depending on the position relative to the laser knife, the bubbles can be "light" or "dark", as well as cast a shadow and reflect a noticeable glare. In the case without adding a fluorescent dye to the flow, but using fluorescent particles for PLIF, the flow pattern is shown in Figure 2d. An optical threshold filter was used to view the position of particles in the flow. In this case, the recognition of bubbles is not required, however, it is possible to obtain information only on the hydrodynamic characteristics of the liquid phase. However, since the liquid flow structure determines the heat transfer, it may be sufficient for some tasks, including the validation of CFD codes.

A series of 10,000 frames for each regime was accrued. Data processing was carried out by "ActualFlow" software developed in IT SB RAS. Firstly, the subtraction of the mean intensity field averaged over the whole sample range were successively applied to enhance the quality of the raw data. Velocity fields were calculated using the iterative cross-correlation algorithm with a continuous window shift and deformation and 75% overlap of the interrogation windows. The initial size of the interrogation window was chosen to be 64 × 64 pixels but it was subsequently reduced to 32 × 32 pixels. The obtained instantaneous velocity vector fields were then validated with the following procedures: peak validation with the threshold of 2.0 and adaptive median 5 × 5 filter.

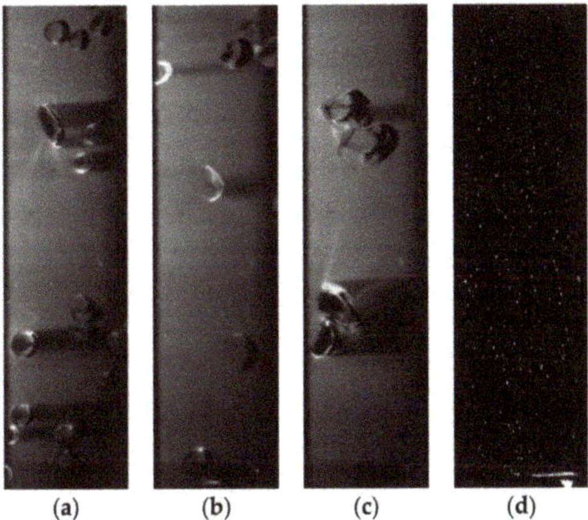

Figure 2. Single frames of the flow at Re = 6600 and β = 0.03: (**a**) "small" bubbles; (**b**) "medium" bubbles; (**c**) "large" bubbles; (**d**) "small" bubbles without Rhodamine in the test fluid.

In order to validate the proposed method, preliminary measurements of the hydrodynamic structure of the flow were carried out using Laser Doppler anemometry system with a measurement uncertainty ±2%. The deviations of the data obtained using the PIV and LDA in a single-phase flow did not exceed 3%.

3. Mathematical Model and Method of Numerical Realization

3.1. D RANS + SMC "In-House" Numerical Code

The flow and heat transfer in the bubbly flow is modeled using the Eulerian approach [26]. It treats the dispersed phase as a continuous medium with properties analogous to those of liquid and this method is widely used for the simulation of turbulent bubbly flow with and without bubble break-up and coalescence processes in various types of flow geometries. The Eulerian approach is based on kinetic equations for a one-point probability density function [27]. The 2D steady-state, incompressible RANS model [12] is used for modeling of turbulent bubbly flow in the BFS. The set of governing equations consists of continuity, two-momentum and energy equations taking into account the effect of bubble presence on the transport processes in the carrier fluid phase. The effect of bubbles on the mean and fluctuational characteristics of turbulent carrier fluid flow is determined by drag, gravity lift, virtual mass, wall lubrication forces, turbulent transport and turbulent diffusion (so called two-way coupling) [12].

The system of 2D RANS equations for the two-phase bubbly flow is given as [23,28]

$$
\begin{gathered}
\nabla \times (\Phi_l \rho U) = 0 \\
\nabla \times (\Phi_l \rho U U) = \Phi_l(-\nabla P + \rho g) + \nabla \times (\Phi_l \rho \tau) - \nabla \times \left(\Phi_l \rho \langle u' u' \rangle + \sum_{k=1}^{K} \sigma_k^{BI} \right) + \\
+(P - P_{in}) \nabla \times \Phi_l + \sum_{k=1}^{K} M_{lk} \\
\nabla \times (\Phi_l \rho C_P U T) = \nabla \times \Phi_l \nabla [(\lambda T)] - \nabla \times (\Phi_l \rho C_P \langle u\theta \rangle) + \sum_{k=1}^{K} \frac{6 h_k (T - T_{bk})}{d_k} + \\
+ \sum_{k=1}^{K} \rho_{bk} C_{Pb,k} g_{ut,k} \langle u\theta \rangle \nabla \times \Phi_l \\
\Phi_l + \Phi_b = \Phi \equiv 1
\end{gathered} \tag{1}
$$

Hereinafter, the index k is partially omitted; Φ_l and Φ_b are the volume fractions of the liquid phase and bubbles, respectively; K is the number of groups of bubbles; P and $P_{in} \approx P_b$ are the pressures in the liquid phase and on the surface of bubble [1,29], respectively; σ^{BI} is the influence of the kth group of bubbles on the tensor of averaged Reynolds stresses in the liquid phase [1,27]; τ, $\langle u'u' \rangle$ and $\langle u\theta \rangle$ are the tensors of viscous stress and Reynolds stress and turbulent heat flow in the carrier phase, respectively; $M_l = \sum_{k=1}^{K} M_{lk} = -M_b$ interfacial term [23,27]; g is the gravity acceleration; $g_{ut,k}$ is the coefficient of involvement of gas bubbles of the kth fraction in the thermal fluctuation movement of the carrier fluid [27]; $\tau_{\Theta k}$ is the thermal relaxation time. The right-hand side of the momentum equation includes the pressure gradient in carrier phase, gravity, the viscous stress, Reynolds stresses in bubbly flow, interfacial pressure gradient and momentum exchange between carrier fluid and gas phases that arise from the actions from interfacial forces.

The turbulence of the carrier fluid is predicted using the Reynolds stress model (second moment closure—SMC) [30] taking the effect of gas bubbles on fluid phase turbulence [31].

$$\begin{gathered}\nabla(\Phi_l \rho U \langle u'u' \rangle) = \nabla \times (\Phi_l D_K) + \Phi_l (P_K + \phi - \varepsilon) + S_K \\ \nabla \times (\Phi_l \rho U \varepsilon) = \nabla \times (\Phi_l D_\varepsilon) + \Phi_l (P_\varepsilon - \varepsilon) + S_\varepsilon \\ \chi - L_T^2 \nabla^2 \chi = \tfrac{1}{\varepsilon T_T}\end{gathered} \tag{2}$$

Here $\phi_{ij} = (1 - k\chi)\phi_{ij}^W + k\chi\phi_{ij}^H$ is the velocity-pressure-gradient correlation, well known as the pressure term [30], χ is the blending coefficient, which goes from zero at the wall to
8 unity far from the wall [27]; ϕ_{ij}^H is "homogeneous" part (valid away from the wall), and ϕ_{ij}^W is "inhomogeneous" part (valid in the wall region) and $L_T = 0.45\, max\left[k^{3/2}/\varepsilon;\ 80(\nu/\varepsilon)^{3/4}\right]$ is the turbulent macro-scale length. Last terms S_K and S_ε in the right part of the turbulence model take into account the effect of the dispersed phase on transport processes [31].

The dispersed phase is modeled by the Eulerian two-fluid approach. The system of axisymmetric mean transport equations for the kth fraction of bubbles has the form [23,28] (to simplify the equations the index k is partially omitted).

$$\begin{gathered}\tfrac{\partial \rho_b}{\partial t} + \nabla \times (\alpha_{bk}\rho_{bk}U_{bk}) = 0 \\ \tfrac{D(\alpha_{bk}\rho_{bk}U_{bk})}{Dt} = \alpha_{bk}\left(\tfrac{D_{bk}}{\tau_k}\nabla \times (\alpha_{bk}\rho_{bk})\right) - \nabla \times (\alpha_{bk}\rho_{bk}E\langle u'u' \rangle) + \sum_{k=1}^{K} M_{bk} \\ \tfrac{D(\alpha_{bk}\rho_{bk}T_{bk})}{Dt} = -\tfrac{1}{\tau_{\Theta k}}\nabla \times \left(\rho_{bk}D_{bk}^{\Theta}\alpha_{bk}\right) - \alpha_{bk}(T - T_{bk})\tfrac{\rho_{bk}}{\tau_{\Theta k}}\rho_{bk} = \alpha_{bk}P_b/(R_b T_{bk})\end{gathered} \tag{3}$$

The method of δ-approximation of [19,20] is used for modeling the polydispersity of the bubbly flow. The break-up and coalescence processes are predicted according to the model of [23,28] and they are described in detail in [32].

The governing transport equations for both gas (1) and dispersed (2) phases, and the turbulence model (3) are solved using a control second-order upwind finite control volumes approach on a staggered grid. The convective terms are discretized using the third order QUICK algorithm and the diffusion terms are numerically solved employing the second-order central difference scheme. The SIMPLEC scheme is used for the coupling of velocity and pressure. The basic grid with 256 × 124 control volumes (CVs) along the longitudinal and transverse directions is used for all numerical simulations. The time derivatives are discretized using the implicit Euler scheme of the first-order accuracy. The components of Reynolds stresses are determined at the same points along the faces of the control volume as the corresponding components of the mean fluid velocity [33]. Grid convergence is verified for three grid sizes is used for the "in-house" code: coarse grid: 128 × 62, basic grid 256 × 124 and fine grid: 400 × 180 control volumes (see Figure 3a).

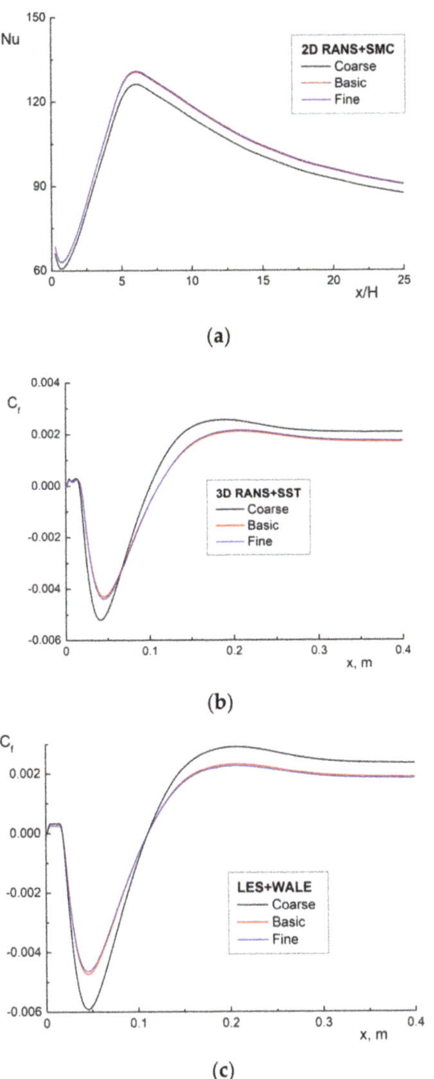

Figure 3. Grid convergence test of 2D RANS + SMC "in-house" code for the bubbly flow at $T_W = 313$ K, $T_1 = T_{b1} = 293$ K, $\beta = 5\%$, $d_1 = 3$ mm (**a**), single-phase fluid flow 3D RANS+SST (**b**) and LES+WALE (**c**) in a duct with sudden expansion. $\text{Re}_H = 6600$, ER = 2.5.

3.2. D RANS + SST and LES (Ansys CFD Package)

Three-dimensional RANS and LES predictions for the single-phase fluid flow behind the single-side BFS are performed in the paper using the commercial CFD package Ansys Fluent 2020R2 [34]. Ansys CFD package was used for the additional validation of the "in-house" RANS + SMC numerical code. 3D RANS simulations by Fluent are employed Menter's k-ω SST model [14] and LES with wall-adapting local eddy-viscosity (WALE) subgrid-scale (SGS) model [35]. The Menter's k-ω SST model is one of the most popular isotropic two-equation turbulence models for simulation of various turbulent flows. The WALE SGS model is one of a major SGS model for flow simulations in complicated geometries.

The Fluent 3D RANS basic grid consist of 256 × 100 × 100 = 2.56 Mio CVs. Grid convergence is verified for three grid sizes: coarse grid: 150 × 50 × 50 CVs, basic grid: 300 × 150 × 100 and fine grid: 450 × 184 × 150 (see Figure 3b). The control volume method for discretization of the governing equations is used. The computational grid is non-uniform, and more refined grid is performed to all wall surfaces. The grid resolution is chosen optimal after preliminary simulations, and it provides the condition for the position of the first cell from the wall $y_+ < 1$. The profiles of the velocity, turbulent kinetic energy and the rate of their dissipation are set at the entrance to the duct with a backward-facing step. They are simulated in the upstream section with the length $5H$ before the section of flow separation. The wall functions are not used, i.e., the predictions are carried out directly to solid surfaces. Boundary conditions at the duct entrance profiles—velocity, turbulent kinetic energy and their dissipation rate—are set after the preliminary simulations of the flow in the upstream section. The temperature of the incoming water flow is constant T_1 = 298 K, and the change in the thermophysical properties is neglected in the predictions.

The LES simulations are conducted using the basic grid with 490 × 124 × 30 = 1.82 Mio CVs. Grid convergence is verified for three grid sizes: coarse grid: 250 × 62 × 20 CVs, basic grid: 490 × 124 × 30 and fine grid: 650 × 190 × 45 (see Figure 3c). It should be noted that the periodic boundary condition in z (duct width) direction for the LES method were used. The governing equations are solved numerically using a pressure-based non-iterative time-advancement (NITA-FS) solver with Courant-Friedrichs-Lewy condition CFL < 0.5. The second-order central scheme is used to discretize the convective terms. To predict the flow in the upstream section, the 3D RANS method was conducted. LES modeling is carried out only for a duct with a single-side backward-facing step. The grid resolution is chosen optimal after preliminary simulations, and it provides the condition for the position of the first cell from the wall $y_+ < 1$. The temperature of the incoming water flow was constant T_1 = 298 K, and the change in the thermophysical properties was neglected in the calculations. The boundary conditions on the wall behind the step is T_W = const = 313 K. The end wall of the step is not heated.

Differences in the value of local Nusselt number Nu calculated for the bubbly flow is less than 0.1%. The maximum error $e_{\max}$ is defined as: $e\ \underset{i=1,\ N}{max}\left|Nu_i^n - Nu_i^{n-1}\right|^{-6}{}_{max}$, where N is the total number of CVs in corresponding direction, the subscript i is the specific CV and the superscript n is the iteration level. The computational grid is nonuniform both in the streamwise and transverse directions. A more refined grid is applied in the recirculation region and in the zones of flow detachment and reattachment and in the inlet region of the duct. The coordinate transformation is suitable for such a two-dimensional problem:

$$\Delta\psi_j = K \times \Delta\psi_{j\text{-}1},$$

where $\Delta\psi_j$ and $\Delta\psi_{j\text{-}1}$ are the current and previous steps of the grid in the axial or radial directions and K = 1.05 (longitudinal direction) and K = 1.03 (transverse direction). The first cell is located at a distance $y_+ = yU_*/\nu$ = 0.3–0.5 from the wall surfaces, where U_* is the friction velocity obtained for the single-phase flow in the inlet of the duct (other parameters being identical). At least 10 CVs were generated to ensure resolution of the mean velocity field and turbulence quantities in the viscosity-affected near-wall region ($y_+ < 10$). The time step was Δt = 0.1 ms. The Courant number, which is a necessary condition for the stability of numerical solution of differential equations in partial derivatives, does not exceed 1 for all simulations.

At the inlet streamwise and transverse components, temperatures of the phases and turbulence levels are uniform. No-slip conditions are set on the wall surface for the carrier fluid phase. At the outlet edge, the computational domain condition $\partial\theta/\partial x = 0$ is set for all variables.

4. Comparison with Experimental and Numerical Results

4.1. Single-Phase Flow in the BFS

At the first stage, comparisons were carried out with the measurement data for a single-phase gas duct flow behind a single-side backward-facing step. The well-known experimental results [36] were used for comparisons on the heat transfer. Additionally, comparisons were performed with the semi-empirical correlation data [37] for minimal wall friction coefficient distribution.

The measured [36] and authors' predicted distributions of Stanton number St along the streamwise coordinate are shown in Figure 4a. The measurements were performed in duct with single-side BFS with dimensions: height 188 mm after sudden expansion and width was 500 mm. The step height was $H = 38$ mm and expansion ratio was ER = 1.25. The carrier fluid was air with temperature $T_1 = 298$ K. The mean upstream fluid velocity $U_{m1} = 11$ m/s and Reynolds number was $\mathrm{Re}_H = 2.8 \times 10^4$. The Stanton number is calculated using the expression: $\mathrm{St} = h/(\rho C_p U_{m1})$. Here h is the heat transfer coefficient, ρ is the density of the carrier fluid flow and C_p is the specific heat at constant pressure.

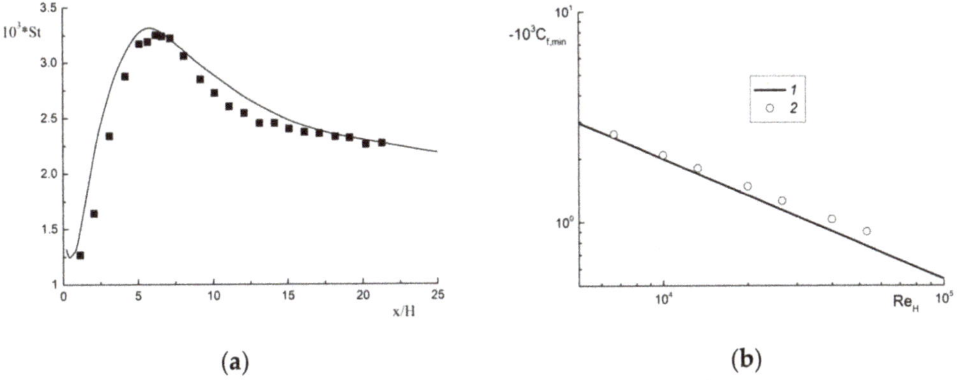

Figure 4. Heat transfer coefficient distributions along the streamwise coordinate (**a**) and minimal wall friction values in the reverse zone vs. Reynolds number (**b**). (**a**): Symbols are the experiments of [36], line is our RANS + SMC "in-house" code simulations. (**b**): 1 is the semi-empirical correlation [37], 2 is our RANS+SMC "in-house" code simulations.

The distributions of the minimum value of the wall friction coefficient $C_{f,\,\min}$ in the separation region for single-phase fluid flow vs. the Reynolds number Re_H based on the step height are given in Figure 4b. The measurements of [37] were performed in the duct with single-side backward-facing step with dimensions: length was 1600 mm, height 70 mm after sudden expansion and width was 220 mm. The step height was $H = 20$ mm and expansion ratio was ER = 1.4. The carrier fluid was water with temperature $T_1 = 298$ K. The mean upstream fluid velocity $U_{m1} = 0.024$–0.24 m/s and Reynolds number varied $\mathrm{Re}_H = (0.12$—$1.22) \times 10^4$. The wall shear stress was measured using the electrodiffusion method. Line *1* is the semi-empirical correlation [37] for calculating the wall friction $C_{f,\min} = -0.38 Re_H{}^{-0.57}$, and points (2) represent the author's predictions. The strong dependency of minimal values of wall friction in the reverse zone on the Reynolds numbers was obtained as well as in experiments [37] and our numerical predictions. Good agreement was obtained between the results [36,37] and the data of authors' numerical predictions.

4.2. Two-Phase Bubbly Flow in Round Vertical Pipe

Comparison of experimental [38] and authors' predicted data on the distributions of carrier fluid and gas bubbles velocities over the round pipe cross-section is shown in Figure 5a. The mean axial velocities of both carrier fluid (line 2) and gas (line 3) phases in a two-phase flow is always higher than in a single-phase flow (line 1). The velocity profile of gas bubbles (line 3) over the cross section of the pipe has a flatter form in comparison with the velocity of the fluid (line 2). The mean axial velocity of the bubbles in the upward flow is higher than that one for the liquid phase due to the action of the Archimedes force.

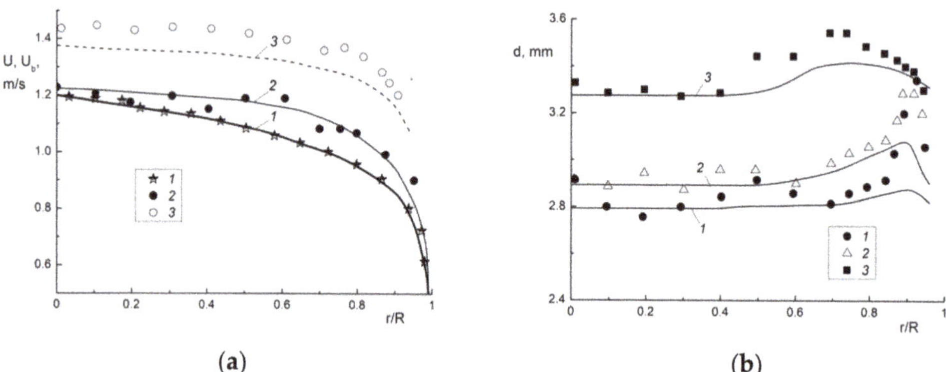

Figure 5. Radial profiles of the carrier fluid and gas bubbles (**a**) and mean Sauter gas bubbles diameters (**b**). β = 4.6%, Re = $U_{m1}2R/\nu$ = 4.9 × 10^4, $x/(2R)$ = 53.5, $2R$ = 50.8 mm, U_{m1} = 0.986 m/s, d_1 = 2.4 mm. Symbols are the measurements of [38], curves are the authors predictions. (**a**): 1, 2 are the carrier fluid axial velocities in the single-phase and two-phase bubbly flows respectively; *3* is the gas bubbles axial velocity. (**b**): 1—J_b = 0.0473 m/s, β = 4.6%, d_1 = 2.4 mm; 2—0.113 m/s, 10.3%, 2.5 mm; 3—0.242 m/s, 19.7%, 2.8 mm.

The radial profiles of the mean Sauter bubble diameter along the pipe radius are shown in Figure 5b. An increase in the gas volumetric flow rate ratios causes an increase in the size of bubbles over the pipe cross-section. The bubble size in the near-wall region is slightly higher than the corresponding value for the core part of a two-phase pipe flow. This conclusion is typical both for measurements [36] and for authors' numerical simulations. This due to the fact that small bubbles migrate to the wall due to the action of the lift and turbulent migration forces and they accumulate near the wall. The probability of their coalescence is increased in this case.

4.3. Two-Phase Bubbly Flow in the BFS

The numerical model was validated against comparison with experimental results for the bubbly flow in a vertical upward pipe with sudden expansion. These results were presented in our previous papers [12,23] and they are not presented here only for brevity. We have carried out the comparison with measurements of [39] and LES of [40] for the isothermal flow regime. The mean-mass velocity was U_{m1} = 1.78 m/s, which corresponds to the Reynolds number Re_H = 1.1 × 10^5. The pipe diameter, before the abrupt, was $2R_1$ = 50 mm, while after expansion, $2R_2$ = 100 mm, which corresponded to the step height of H = 25 mm, and ER = $(R_2/R_1)^2$ = 4 and the inlet bubble Sauter diameter was d_1 = 2 mm. The authors results agree well with experiments and LES data. We have conducted the comparison with present measurements for the mean fluid axial velocity profiles and heat transfer enhancement [23]. The maximal difference between our and other authors measured results and our predictions is up to 20%.

4.4. Two-Phase Bubbly Flow behind a Backward-Facing Step. Our Measurements and Numerical Simulation and Their Discussion

Based on a successful comparison of the results obtained using different methods, studies of the local structure of the two-phase bubbly flow in a BFS are carried out. Symbols and curves are the authors' measurements and predictions, respectively. The transverse profiles of measured and predicted mean streamwise fluid velocities at the Re_H = 6600 in the single-phase fluid flow (lines 1) and at gas volumetric flow rate ratio β = 3% (lines 2) along the duct length are shown in Figure 6. The carrier fluid velocity was normalized to the maximum value of liquid velocity. The point of flow reattachment in the single-phase fluid flow is located at a distance of (7–8)H, which is in good agreement with the literature data [41,42]. The following characteristic features of the bubbly flow can be distinguished. Immediately behind the step, the flow recirculation zone is observed. The gas bubbles badly penetrate this region and the effect of gas bubbles on the flow is rather weak in this area. The position of the flow reattachment point shifts towards the step for the two-phase flow and, in general, the establishment of a profile in the channel after the expansion occurs faster in a two-phase flow than in a single-phase flow, which was previously reported, for example, in [23]. Our numerical results capture the main features of the bubbly flow behind the BFS described above. The main difference between the experimental and numerical results for the two-phase bubbly is obtained in the near-wall zone in the flow relaxation region.

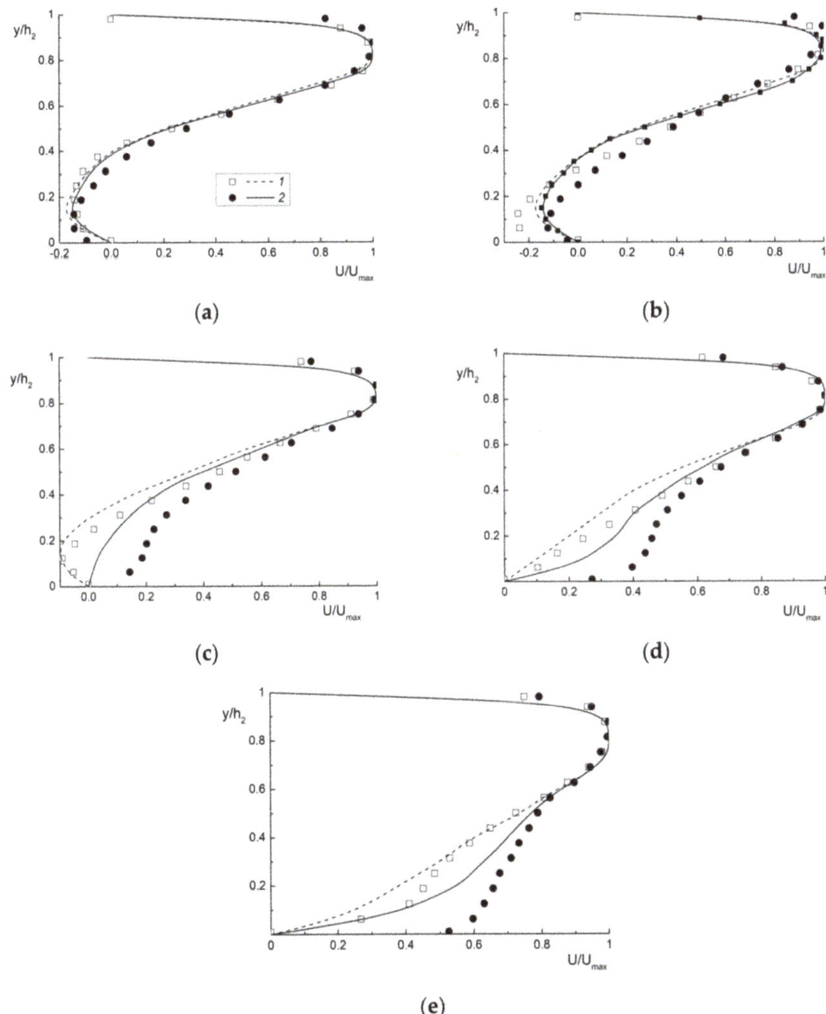

Figure 6. The transverse profiles of the carrier phase mean streamwise velocity along the duct length. Symbols and curves are the authors' measurements and predictions respectively. Re_H = 6600, ER = 2.5, β = 3%. 1—single-phase flow (β = 0), 2—β = 3%. (**a**)—x/H = 2, (**b**)—4, (**c**)—6, (**d**)—8, (**e**)—10.

5. Numerical Results and Its Discussion

5.1. Single-Phase Fluid Flow in a Backward-Facing Step Flow

The distribution of mean streamwise fluid velocity component at various distances from the BFC is presented in Figure 7. The measurements are conducted using the PIV. The 3D RANS + SST predicted contours of the longitudinal velocity component and streamlines at a few stations along the z axis are shown in Figure 8a. The computations are carried out using the commercial CFD package Ansys Fluent 2020R2. This figure clearly shows the three-dimensional flow pattern in the recirculation area. The length of the recirculation zone is maximum in the central section (z/H = 8.3). The length of the recirculation zone behind the backward-facing step increases with distance from the side walls of the duct. When flowing around the backward-facing step, the flow detaches the sharp edge, and a separation bubble is formed. A long recirculation zone and, characteristic of it, a secondary vortex is formed behind the edge. It has the opposite direction of flow rotation to the main one, and the rotation velocity in it is relatively low, which leads to the formation of a stagnant zone in this region.

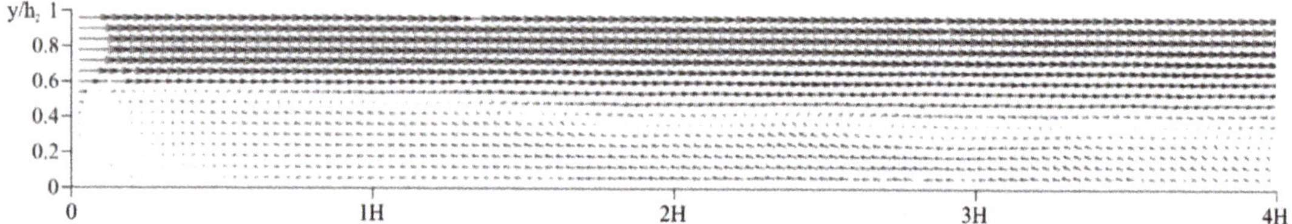

Figure 7. Velocity distribution on the channel centerline.

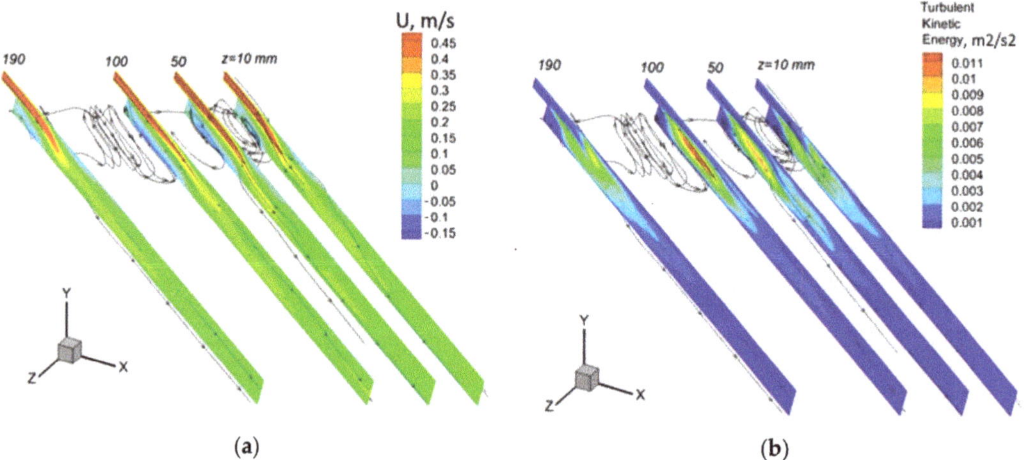

Figure 8. The 3D RANS + SST predicted contours of the longitudinal velocity component and streamlines (**a**) and turbulent kinetic energy (**b**) at the z = 10, 50, 100, 190 mm (z/H = 0.8, 4.2, 8.3 and 15.8).

The change in the turbulent kinetic energy behind the step in various sections along the z axis is presented in Figure 8b. A turbulization of the flow occurs when flowing in the BFS due to additional vortex formation. The largest values of TKE are located in the region of the mixing shear layer, and their minimum values are observed in the region of the secondary vortex. The turbulent kinetic energy increases toward the center duct. The largest values of turbulent kinetic energy in the region are observed in the central section.

The transverse profiles of single-phase flow mean streamwise velocity behind a single-side backward-facing step are shown in Figure 9. Here symbols and curves are the authors' measurements and predictions, respectively. The first three sections are located in the recirculation region, the fourth station is set close to the reattachment point and last two sections are set in the flow relaxation zone. It was obtained a qualitatively agreement between authors' measurements and LES and RANS results.

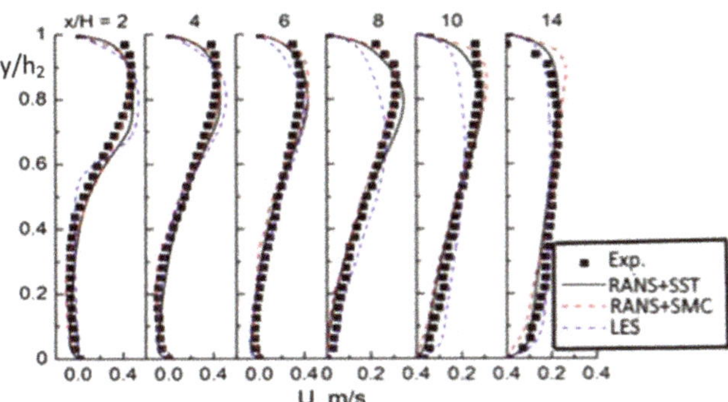

Figure 9. The transverse profiles of single-phase mean longitudinal velocity behind a backward-facing step. Symbols and curves are the authors' measurements and predictions, respectively. $Re_H = 6600$, ER = 2.5.

5.2. *Upward Bubbly Flow in a Backward-Facing Step Flow*

Numerical simulations are carried out for a fluid flow of a water and polydispersed air bubbles at atmospheric pressure and all these results given in this subsection are obtained using the "in-house" 2D RANS + SMC model. The upward bubbly flow in the duct with a single-side backward-facing step is considered (see Figure 1b). The Reynolds number $Re = U_{m1}H/\nu = 6600$, and the mean-mass velocity at the inlet $U_{m1} = 0.55$ m/s. The height of the duct before separation $h_1 = 8$ mm, height of the duct after separation $h_2 = 20$ mm, and the step height $H = 12$ mm, the expansion ratio ER = $(H + h_1)/h_1 = h_2/h_1 = 2.5$, the wall temperature T_W = const = 313 K and the initial temperatures of carrier liquid and gas bubbles $T_1 = T_{b1} = 293$ K. In the inlet section, the mean phase velocities have the same value. The initial gas volumetric flow rate ratios are varied $\beta = 0$–10% and the mean initial Sauter diameter of air bubbles $d_1 = 3$ mm. All simulations are performed for four monodispersed δ-functions (modes) of air bubbles at the inlet cross-section: 1—$d/d_1 = 0.33$, 2—0.66, 3—1, and 4—1.33. The volume fraction of each fraction is $\Phi_1(1\text{ mm}) = 0.031\Phi$, $\Phi_2(2\text{ mm}) = 0.12\Phi$, $\Phi_3(3\text{ mm}) = 0.6\Phi$, and $\Phi_4(4\text{ mm}) = 0.15\Phi$, where Φ is the total volume fraction of gas bubbles. Authors did not carry out the measurements of the distributions of the diameters of gas bubbles at the inlet cross-section. Therefore, the choice of just such values of gas bubbles diameters and their volume fractions is based on our preliminary numerical predictions. There is no steam generation on wall surface. The same assumptions were used in our previous recent numerical simulations [12,13,29], but they may be important in other thermal boundary conditions on the wall [43].

In Figure 10 the transverse profiles of the total local void fraction (a) and total bubble diameter (b) of all monodispersed four modes are shown along the duct length in the upward bubbly flow behind the backward-facing step. The so-called "top-hat" distribution [44] of the local void fraction is obtained at the inlet (line 1) (the cross-section of sudden expansion). This is one of the most typical distributions of void fraction in a vertical upward bubbly flow for a low gas bubble concentration. The obvious maximum of void fraction in the near-wall regions is typical for this type of void fraction distribution [44]. Further, there is a significant change in the predicted distributions of the local void fraction over the duct cross-section. It can be explained by a significant increase in the two-phase flow cross-section behind the BFS (see Figure 10a). The "right" local maximum of the void fraction is also located near the "right" wall of the duct, but its value is noticeably less than for the inlet section (lines 2–5). The "left" local maximum of the void fraction becomes much less noticeable along the duct length by bubbles scattering and diffusion across the duct cross-section. In the area of the recirculation zone of the two-phase flow (curves 2–4), an extremely small number of gas bubbles are obtained. It can be explained by their relatively large initial diameter. It was recently shown in our previous work [12] that large bubbles practically do not penetrate the recirculation region in a pipe with sudden expansion and are observed mainly in the mixing layer and in the flow core. The bubbles distribution after the reattachment point is characterized by the presence of bubbles over the entire all cross section of the duct by the effect forces acting on a bubble in a transverse direction and low maximum in the near-wall zone of the duct (line 5). The profile of local void fraction at $x/H = 10$ has almost uniform distribution. The profiles of the local void fraction are characterized by a zero-value close to the duct wall. This is explained by the fact under the action of the wall force a bubble cannot approach a wall. The profiles of dimensionless bubble diameter of all monodispersed four δ-functions are presented in Figure 10b. The maximum bubbles size is obtained in the near-wall zone of the duct. The bubbles diameter in the recirculation region is less than that one in the flow core.

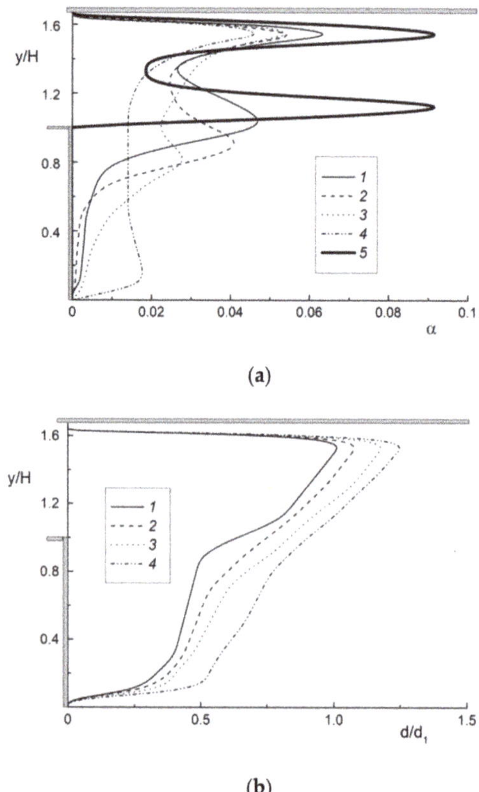

Figure 10. The transverse profiles of total local void fraction (**a**) and bubble diameter (**b**) along the duct length in the upward bubbly flow behind the backward-facing step. $Re_H = 6600$, ER = 2.5, $\beta = 3\%$. (**a**): 1—$x/H = 0$, 2—2, 3—4, 4—6, 5—10; (**b**): 1—$x/H = 2$, 2—4, 3—6, 4—10.

The "in-house" 2D RANS + SMC numerical results for the sum (total) of all four fractions of air bubbles were shown in Figure 11. To better understand the complicated and coupled break-up and coalescence processes in bubbly turbulent flows, the results of local void fractions and bubble diameter distributions obtained for four various bubble fractions are important, and these results are presented in Figure 11. The cross-section $x/H = 4$ is located in the area of flow recirculation. The void fraction profiles of the smallest (line 1) and middle (curve 2) bubbles have only one local maximum located close to the duct wall. Only the bubbles of the smallest size (line 1) are observed in the recirculation zone of the duct (see Figure 11a). The void fraction profiles of large bubbles (curves 3 and 4) have two local maxima and are set close to the duct wall and in the shear mixing layer. Two obvious local maxima are revealed in the distribution of void fraction of all modes (line 5) too.

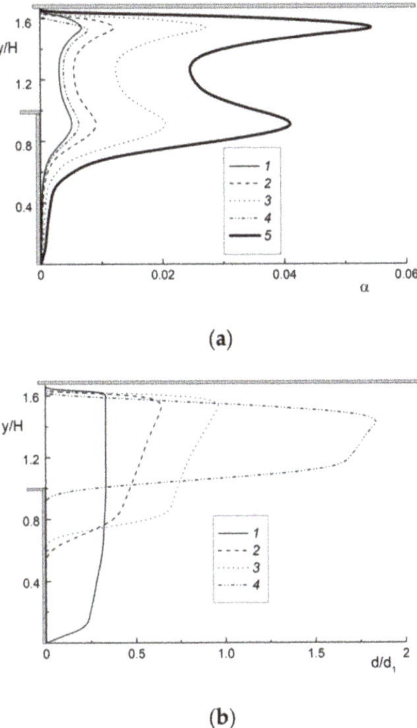

Figure 11. The transverse profiles of local void fractions (**a**) and bubble diameter (**b**) for various bubble size modes. Re_H = 6600, ER = 2.5, β = 3%, x/H = 4. (**a**): 1—d/d_1 = 0–0.33, 2—0.33–0.66, 3—0.66–1, 4—>1, 5—sum of all modes; (**b**): 1—d = 0–0.33, 2—0.33–0.66, 3—0.66–1, 4—>1.

The distribution of all fractions of bubble diameters across the duct cross-section is practically uniform (see Figure 11b). Bubbles of the smallest investigated diameter $d/d_1 \leq 0.33$ (line 1) are found over the entire section of the channel, both in the flow recirculation zone and in the flow core. The smallest air bubbles can come closer to the duct wall than the larger bubbles due to the effect of transverse forces (lift, turbulent migration, turbulent diffusion and wall force). The largest bubbles $d/d_1 \geq 1$ (line 5) are located mainly in the near-wall part of the channel, which confirms the data of our numerical predictions presented in Figure 11a,b. Bubbles of the other two monodispersed fractions (d/d_1 = 0.33–0.66, 0.66–1, lines 2 and 3) also practically do not penetrate the flow separation region.

The effect of the volumetric flow rate gas content on the profiles of the longitudinal liquid velocity is shown in Figure 12. The carrier fluid (water) velocity in the two-phase bubbly flow (curves 2 and 3) is slightly higher that one in the single-phase flow (line 1). This effect increases with an increase in the concentration of air bubbles.

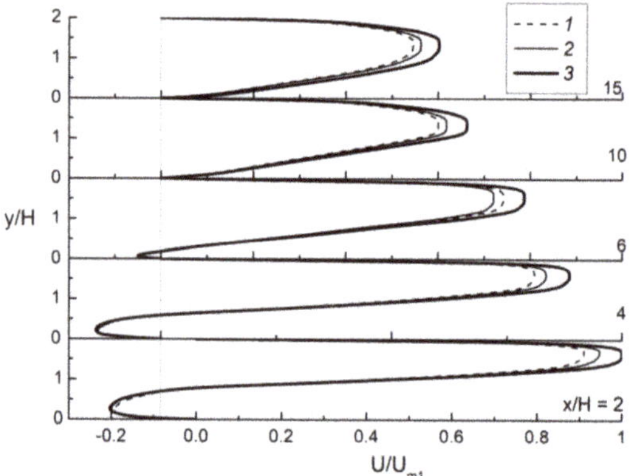

Figure 12. Profiles of mean axial fluid velocity in bubbly flow in the backward-facing step. 1—β = 0 (single-phase fluid flow, other conditions being identical), 2—2%, 3—5%.

The effect of the gas volumetric flow rate ratios (a) and gas bubbles diameter at the inlet (b) on the carrier fluid phase turbulent kinetic energy is shown in Figure 13. Carrier fluid phase turbulence level for a 2D flow is estimated using the expression: $2k = \langle u_i u_i \rangle = u'^2 + v'^2 + w'^2 \approx 1.5(u'^2 + v'^2)$. The maximum of TKE of the fluid in the two-phase flow is located in the shear mixing layer. The significant enhancement of the carrier fluid phase turbulence level in the two-phase flow (up to 40% at β = 10%) is shown in comparison with the single-phase flow (bold solid curves). The increase of initial bubble size causes a significant increase in the turbulent kinetic energy in the bubbly flow (up to 25% at d_1 = 3 mm). The additional production of carrier fluid phase turbulence is explained by vortex formation upon streamlining of the gas bubbles by the carrier fluid flow. The profiles of TKE at a small value of β agree qualitatively with those for the case of one-phase flow in a duct with single-side BFS. The same tendencies were obtained in our previous works [12,23] for a pipe with sudden expansion.

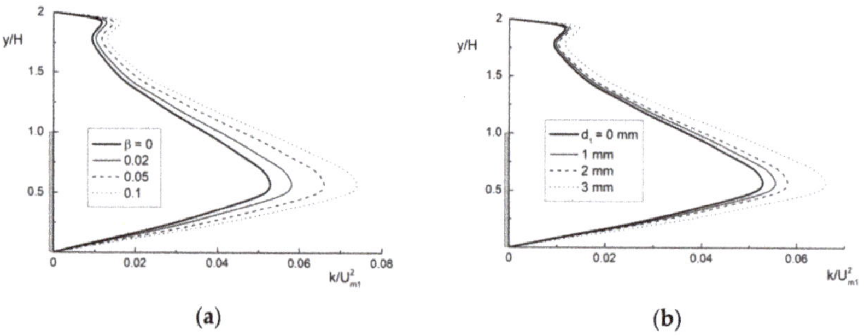

Figure 13. The effect of the gas volumetric flow rate ratios (**a**) and gas bubbles diameter at the inlet (**b**) on the carrier fluid phase turbulence.

In Figure 14 the effect of the addition of the gas bubbles on heat transfer from the wall to the two-phase bubbly flow. The local convective Nusselt number at constant wall temperature is estimated:

$$\mathrm{Nu} = -(\partial T/\partial y)_W H/(T_W - T_m)$$

where $(\partial T/\partial y)_W$ is the gradient of the fluid temperature on the wall, T_W is the wall temperature and T_m is the mean temperature of the carrier fluid (water) in this section. Lines 1 are Nusslet number in the fully developed single-phase duct flow and dashed curves 2 are the predictions of heat transfer in the single-phase duct flow with sudden expansion for the fluid (water) flow with other conditions being identical.

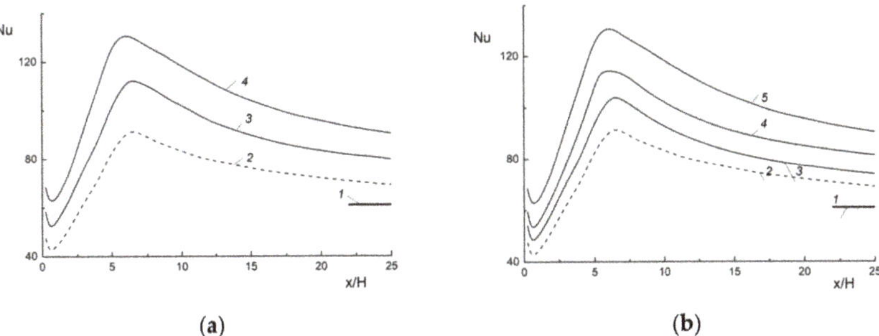

Figure 14. Heat transfer in the bubbly flow behind the backward-facing step along the streamwise coordinate. (**a**): d_1 = 3 mm, 1—Nusslet number value in the fully developed single-phase duct flow, 2—β = 0 (single-phase fluid flow, other conditions being identical), 3—2%, 4—5%; (**b**): β = 5%, 1—Nusslet number value in the fully developed single-phase duct flow, 2—d_1 = 0 (single-phase fluid flow, other conditions being identical), 3—0.5 mm, 4—2 mm, 5—3 mm.

A noticeable increase of heat transfer (up to 35% at d_1 = 3 mm in comparison with the single-phase fluid flow) with an increase in the volumetric flow rate ratio β is observed (see Figure 14a). This is explained by an increase in the velocity and temperature gradients and turbulization of the carrier fluid (water) phase in the near-wall region of the duct. The position of the peak of heat transfer shifts upstream and at β = 5% is x_{Nu_max}/H = 4.5, and the length of the recirculation zone approximately coincides with it, x_R/H = 4.8 at β = 5%. The same values for the single-phase fluid flow are x_{Nu_max}/H = 6.8 and x_R/H = 7. The position of the peak of heat transfer rate x_{Nu_max} is close to the position of the flow reattachment point x_R in two-phase bubbly flow. The same tendency was observed in [41,42] for single-phase fluid flows in a BFS. The effect of bubble diameter in the inlet on the Nusselt number distributions along the duct length is given in Figure 14b. The largest increase in heat transfer (approximately 35% at fixed value β = 5%) is characteristic for the largest gas bubbles with the initial diameter d_1 = 3 mm.

The effect of gas volumetric flow rate ratios β on values of minimal wall friction coefficient $C_{f,\min}/C_{f,\min,0}$, maximal heat transfer enhancement ratios $\mathrm{Nu}_{\max}/\mathrm{Nu}_{\max,0}$, recirculation length x_R, maximal values of turbulence modification ratios $k_{\max}$ and position of heat transfer maximum $x_{\max}$ are shown in Figure 15. All these variables are normalized on the value in the single-phase fluid flow and subscript "0" is the in the single-phase flow parameter with other conditions are identical to the two-phase bubbly flow.

It is obtained that the magnitude of minimal wall friction coefficient decreases with the increase in the bubble concentration and for β = 10% the minimal value of wall friction ratio is $C_{f,\min}/C_{f,\min,0} \approx 0.75$. The reduction of minimal value of wall friction ratio is small (up to 10%) for the $\beta \leq 5\%$. Almost linear convective heat transfer enhancement is observed for the range studied of gas volumetric flow rate ratios and for β = 10% the heat transfer augmentation ratio is $\mathrm{Nu}_{\max}/\mathrm{Nu}_{\max,0} \approx 1.75$. The addition of gas bubbles leads to the significant shortening of the recirculation length ratio of bubbly flow. The shortening of the recirculating zone is almost twice in comparison with the case of the single-phase fluid flow. The main reason is the flow turbulization by flowing around the gas bubbles. The maximal value of turbulent kinetic energy modification is up to 50% at β = 10%. It causes the intensification of mixing process. The same trends were obtained for the single-phase separated flows [41,42]. It is well known [41,42] that the position of maximal value of heat transfer is close to the flow reattachment point for the single-phase for both gas (air) and fluid (liquid) flows behind a BFS and a sudden pipe expansion. The addition of relatively large air bubbles shifted the position of maximal value of heat transfer far from the reattachment point for the bubbly flow in the pipe with sudden expansion [12].

In the case of bubbly flow in the backwards-facing step we observe the same trends. The position of the heat transfer maximum is located after the reattachment point.

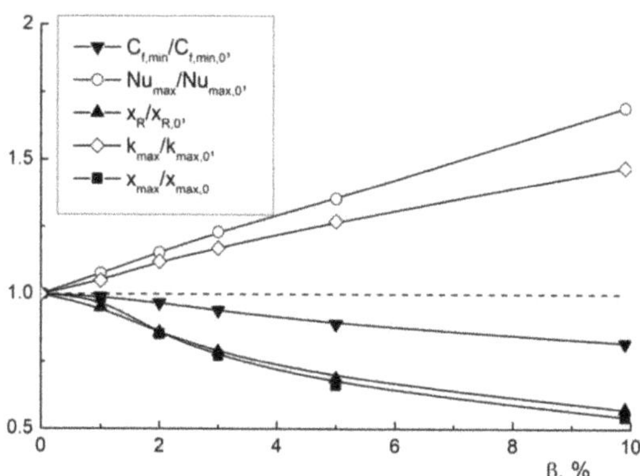

Figure 15. The magnitudes of minimal wall friction coefficient (1), maximal heat transfer enhancement ratios (2), recirculation length (3), maximal values of turbulence modification ratios (4) and position of heat transfer maximum (5) depending on gas volumetric flow rate ratios β. $Re_H = 6600$, ER = 2.5.

6. Conclusions

The PIV/PLIF measurements of mean streamwise flow structure and RANS numerical simulation of the bubbly polydispersed upward duct flow in a backward-facing step are performed. Measurements of the carrier fluid phase velocity are performed using the PIV/PLIF system. The numerical model is based on the Eulerian two-fluid approach. Turbulence of the carrier fluid phase is predicted using the Reynolds stress model.

Experimental study of the characteristics of the motion of gas bubbles behind a single side backward facing step with variations in the volumetric flow rates of carrier fluid and gas have been carried out. The distribution of the mean streamwise gas bubbles and liquid velocities are obtained. It is shown that the bubbles slow down at a distance from the flow detachment cross-section, and it is associated with the slowing down of the liquid velocity due to the increase in the duct cross-section. The bubbles have a more curvilinear trajectory in the separation zone than in the duct before the sudden expansion cross-section. It is shown that bubble clusters can form in the flow relaxation region.

Small bubbles are presented over the entire duct cross-section, and the larger bubbles mainly observed in the shear mixing layer and flow core. The recirculation length in two-phase bubbly flow is shorter (up to twice) than that one in the case single-phase flow. The position of the heat transfer maximum is located after the reattachment point. The effect of the gas volumetric flow rate ratios on the flow and heat transfer in the two-phase flow is numerically studied. The addition of air bubbles results in a significant increase in the heat transfer (up to 75%).

Author Contributions: Conceptualization, P.D.L. and M.A.P.; methodology, T.V.B., A.V.C., P.D.L. and M.A.P.; investigation, T.V.B., I.A.E., D.V.K., P.D.L. and M.A.P.; data curation, M.A.P.; formal analysis, P.D.L. and M.A.P.; writing—original draft preparation, P.D.L. and M.A.P.; writing—review and editing, P.D.L. and M.A.P.; resources, P.D.L. and M.A.P.; project administration, P.D.L. and M.A.P.; All authors have read and agreed to the published version of the manuscript.

Funding: The numerical part of the paper was funded by Ministry of Science and Higher Education of the Russian Federation (mega-grant 075-15-2021-575) and the measurements were performed under IT SB RAS state assignment (Project No. 121032200034-4).

Institutional Review Board Statement: Not applicable.

Informed Consent Statement: Not applicable.

Data Availability Statement: Data are contained within the article.

Conflicts of Interest: The authors declare no conflict of interest.

Nomenclature

$C_f = 2\tau_W/U_1^2$	wall friction coefficient
C_P	heat capacity
d	mean bubble Sauter diameter
h_1	height of the duct before the sudden expansion
h_2	height of the duct after the sudden expansion
H	step height
J	and J_b superficial velocity of carrier fluid (water) and gas bubbles respectively
$2k = \langle u_i u_i \rangle$	turbulent kinetic energy
L	duct length
$\mathrm{Nu} = -(\partial T/\partial y)_W H/(T_W - T_m)$	Nusslet number
$\mathrm{Re}_H = U_{m1}H/\nu$	the Reynolds number
$\mathrm{St} = h/(\rho C_p U_{m1})$	Stanton number
T	temperature
U_{m1}	mean-mass flow velocity
U_*	friction velocity
x	streamwise coordinate
x_R	position of the flow reattachment point
$x_{\mathrm{Nu\_max}}$	position of the peak of heat transfer rate
y	distance normal from the wall
Subscripts	
0	single-phase fluid (water) flow
1	initial condition
W	wall
b	bubble
l	liquid
m	mean-mass
Greek	
Φ	volume fraction
α	void fraction
β	gas volumetric flow rate ratio
ε	dissipation of the turbulent kinetic energy
λ	thermal conductivity
ρ	density
ν	kinematic viscosity
τ_W	wall shear stress
Acronym	
BFS	backward-facing step
CV	control volume
DNS	direct numerical simulation
ER	expansion ratio
LES	large eddy simulation
PIV	particle image velocimetry
PLIF	planar laser induced fluorescence
RANS	Reynolds-averaged Navier-Stokes
ROI	region of interests
SMC	second moment closure
SST	shear stress tensor
TKE	turbulent kinetic energy

References

1. Nigmatulin, R.I. *Dynamics of Multiphase Media*; Hemisphere: New York, NY, USA, 1991; Volume 1.
2. Ishii, M.; Hibiki, T. *Thermo-Fluid Dynamics of Two-Phase Flow*; Springer: Berlin/Heidelberg, Germany, 2011.
3. Aloui, F.; Souhar, M. Experimental study of a two-phase bubbly flow in a flat duct symmetric sudden expansion. Part I: Visulization, pressure and void fraction. *Int. J. Multiph. Flow* **1996**, *22*, 651–665. [CrossRef]
4. Schmidt, J.; Friedel, L. Two-phase pressure change across sudden expansions in duct areas. *Chem. Eng. Commun.* **1996**, *141*, 175–190. [CrossRef]
5. Aloui, F.; Doubliez, L.; Legrand, J.; Souhar, M. Bubbly flow in an axisymmetric sudden expansion: Pressure drop, void fraction, wall shear stress, bubble velocities and sizes. *Exp. Therm. Fluid Sci.* **1999**, *19*, 118–130. [CrossRef]
6. Ahmed, W.H.; Ching, C.Y.; Shoukri, M. Development of two-phase flow downstream of a horizontal sudden expansion. *Int. J. Heat Fluid Flow* **2008**, *29*, 194–206. [CrossRef]
7. Balakhrisna, T.; Ghosh, S.; Das, G.; Das, P.K. Oil–water flows through sudden contraction and expansion in a horizontal pipe—phase distribution and pressure drop. *Int. J. Multiph. Flow* **2010**, *36*, 13–24. [CrossRef]
8. Arabi, A.; Salhi, Y.; Si-Ahmed, E.K.; Legrand, J. Influence of a sudden expansion on slug flow characteristics in a horizontal two-phase flow: A pressure drop fluctuations analysis. *Meccanica* **2018**, *53*, 3321–3338. [CrossRef]
9. Zhang, D.W.; Goharzadeh, A. Effect of sudden expansion on two-phase flow in a horizontal pipe fluid dynamics. *Fluid Dyn.* **2019**, *54*, 123–136. [CrossRef]
10. Kim, Y.; Park, H. Upward bubbly flows in a square pipe with a sudden expansion: Bubble dispersion and reattachment length. *Int. J. Multiph. Flow* **2019**, *118*, 254–269. [CrossRef]
11. Krepper, E.; Beyer, M.; Frank, T.; Lucas, D.; Prasser, H.-M. CFD modelling of polydispersed bubbly two-phase flow around an obstacle. *Nucl. Eng. Des.* **2009**, *239*, 2372–2381. [CrossRef]
12. Pakhomov, M.A.; Terekhov, V.I. Modeling of the flow patterns and heat transfer in a turbulent bubbly polydispersed flow downstream of a sudden pipe expansion. *Int. J. Heat Mass Transfer* **2016**, *101*, 1251–1262. [CrossRef]
13. Bogatko, T.V.; Pakhomov, M.A. Modeling bubble distribution and heat transfer in polydispersed gas-liquid flow in a backward-facing step. *IOP J. Phys. Conf. Ser.* **2020**, *1677*, 012052. [CrossRef]
14. Menter, F.R. Two-equation eddy-viscosity turbulence models for engineering applications. *AIAA J.* **1994**, *32*, 1598–1605. [CrossRef]
15. Krepper, E.; Lucas, D.; Frank, T.; Prasser, H.-M.; Zwart, P.J. The inhomogeneous MUSIG model for the simulation of polydispersed flows. *Nucl. Eng. Des.* **2008**, *238*, 1690–1702. [CrossRef]
16. Lakehal, D.; Smith, B.L.; Milelli, M. Large-eddy simulation of bubbly turbulent shear flows. *J. Turbul.* **2002**, *3*, 1–21. [CrossRef]
17. Dhotre, M.T.; Niceno, B.; Smith, B.L. Large eddy simulation of a bubble column using dynamic sub-grid scale model. *Chem. Eng. J.* **2008**, *136*, 337–348. [CrossRef]
18. Long, S.; Yang, J.; Huang, X.; Li, G.; Shi, W.; Sommerfeld, M.; Yang, X. Large-eddy simulation of gas-liquid two-phase flow in a bubble column reactor using a modified sub-grid scale model with the consideration of bubble-eddy interaction. *Int. J. Heat Mass Transfer* **2020**, *161*, 120240. [CrossRef]
19. Zaichik, L.I.; Mukin, R.V.; Mukina, L.S.; Strizhov, V.F. Development of a diffusion-inertia model for calculating bubble turbulent flows: Isothermal polydispersed flow in a vertical pipe. *High Temp.* **2012**, *50*, 621–630. [CrossRef]
20. Mukin, R.V. Modeling of bubble coalescence and break-up in turbulent bubbly flow. *Int. J. Multiphase Flow* **2014**, *62*, 52–66. [CrossRef]
21. Lehr, F.; Mewes, D. A transport equation for the interfacial area density applied to bubble columns. *Chem. Eng. Sci.* **2001**, *56*, 1159–1166. [CrossRef]
22. Pour, M.; Nassab, S. Numerical investigation of forced laminar convection flow of nanofluids over a backward facing step under bleeding condition. *J. Mech.* **2012**, *28*, N7–N12. [CrossRef]
23. Lobanov, P.D.; Pakhomov, M.A.; Terekhov, V.I. Experimental and numerical study of the flow and heat transfer in a bubbly turbulent flow in a pipe with sudden expansion. *Energies* **2019**, *12*, 2375. [CrossRef]
24. Poletaev, I.; Tokarev, M.P.; Pervunin, K.S. Bubble patterns recognition using neural networks: Application to the analysis of a two -phase bubbly jet. *Int. J. Multiph. Flow* **2020**, *126*, 103194. [CrossRef]
25. Cerqueira, R.F.L.; Paladino, E.E. Development of a deep learning-based image processing technique for bubble pattern recognition and shape reconstruction in dense bubbly flows. *Chem. Eng. Sci.* **2021**, *230*, 116163. [CrossRef]
26. Drew, D.A. Mathematical modeling of two-phase flow. *Ann. Rev. Fluid Mech.* **1983**, *15*, 261–291. [CrossRef]
27. Zaichik, L.I.; Alipchenkov, V.M. A statistical model for predicting the fluid displaced/added mass and displaced heat capacity effects on transport and heat transfer of arbitrary density particles in turbulent flows. *Int. J. Heat Mass Transfer* **2011**, *54*, 4247–4265. [CrossRef]
28. Pakhomov, M.A.; Terekhov, V.I. Modeling of flow structure, bubble distribution, and heat transfer in polydispersed turbulent bubbly flow using the method of delta function approximation. *J. Eng. Thermophys.* **2019**, *28*, 453–471. [CrossRef]
29. Drew, D.A.; Lahey, R.T., Jr. Application of general constitutive principles to the derivation of multidimensional two-phase flow equations. *Int. J. Multiph. Flow* **1979**, *5*, 243–264. [CrossRef]
30. Fadai-Ghotbi, A.; Manceau, R.; Boree, J. Revisiting URANS computations of the backward-facing step flow using second moment closures. Influence of the numerics. *Flow Turbul. Combust* **2008**, *81*, 395–410. [CrossRef]

31. Lopez de Bertodano, M.; Lee, S.J.; Lahey, R.T., Jr.; Drew, D.A. The prediction of two-phase turbulence and phase distribution using a Reynolds stress model. *ASME J. Fluids Eng.* **1990**, *112*, 107–113. [CrossRef]
32. Nguyen, V.T.; Song, C.-H.; Bae, B.U.; Euh, D.J. Modeling of bubble coalescence and break-up considering turbulent suppression phenomena in bubbly two-phase flow. *Int. J. Multiph. Flow* **2013**, *54*, 31–42. [CrossRef]
33. Hanjalic, K.; Jakirlic, S. Contribution towards the second-moment closure modelling of separating turbulent flows. *Comput. Fluids* **1998**, *27*, 137–156. [CrossRef]
34. Available online: www.ansys.com.
35. Nicoud, F.; Ducros, F. Subgrid-scale stress modelling based on the square of the velocity gradient tensor. *Flow Turbul. Combust* **1999**, *62*, 183–200. [CrossRef]
36. Vogel, J.C.; Eaton, J.K. Combined heat transfer and fluid dynamics measurements downstream of a backward facing step. *ASME J. Heat Transfer* **1985**, *107*, 922–929. [CrossRef]
37. Tihon, J.; Legrand, J.; Legentilhomme, P. Near-wall investigation of backward-facing step flows. *Exp. Fluids* **2001**, *31*, 484–493. [CrossRef]
38. Hibiki, T.; Ishii, M.; Xiao, Z. Axial interfacial area transport of vertical bubbly flows. *Int. J. Heat Mass Transfer* **2001**, *44*, 1869–1888. [CrossRef]
39. Bel Fdhila, R. Analyse Experimental et Modelisation d'un Ecoulement Vertical a Bulles dans un Elargissement Brusque. Ph.D. Thesis, Institut National Polytechnique de Toulouse, Toulouse, France, 1991.
40. Papoulias, D.; Splawski, A.; Vikhanski, A.; Lo, S. Eulerian multiphase predictions of turbulent bubbly flow in a step-channel expansion. In Proceedings of the Nineth International Conference on Multiphase Flow ICMF-2016, Firenze, Italy, 22–27 May 2016.
41. Eaton, J.K.; Johnston, J.P. Review of research on subsonic turbulent flow reattachment. *AIAA J.* **1981**, *19*, 1093–1100. [CrossRef]
42. Chen, L.; Asai, K.; Nonomura, T.; Xi, G.N.; Liu, T.S. A review of backward-facing step (BFS) flow mechanisms, heat transfer and control. *Thermal Sci. Eng. Progr.* **2018**, *6*, 194–216. [CrossRef]
43. Wang, H.G.; Zhang, C.G.; Xiong, H.B. Growth and collapse dynamics of a vapor bubble near or at a wall. *Water* **2021**, *13*, 12. [CrossRef]
44. Kashinsky, O.N.; Chinak, A.V.; Kaipova, E.V. Bubbly gas-liquid flow in an inclined rectangular channel. *Thermophys. Aeromech.* **2003**, *10*, 69–75.

Article

Study of Pressure Drops and Heat Transfer of Nonequilibrial Two-Phase Flows

Aleksandr V. Belyaev [1,*], Alexey V. Dedov [1], Ilya I. Krapivin [1], Aleksander N. Varava [1], Peixue Jiang [2] and Ruina Xu [2]

1 Department of General Physics and Nuclear Fusion, The National Research University "MPEI", Krasnokazarmennaya 14, 111250 Moscow, Russia; dedovav@mpei.ru (A.V.D.); KrapivinII@mpei.ru (I.I.K.); VaravaAN@mpei.ru (A.N.V.)

2 Department of Thermal Engineering, Tsinghua University, Haidian District, Beijing 100084, China; jiangpx@tsinghua.edu.cn (P.J.); ruinaxu@tsinghua.edu.cn (R.X.)

* Correspondence: BeliayevAVL@mpei.ru

Abstract: Currently, there are no universal methods for calculating the heat transfer and pressure drop for a wide range of two-phase flow parameters in mini-channels due to changes in the void fraction and flow regime. Many experimental studies have been carried out, and narrow-range calculation methods have been developed. With increasing pressure, it becomes possible to expand the range of parameters for applying reliable calculation methods as a result of changes in the flow regime. This paper provides an overview of methods for calculating the pressure drops and heat transfer of two-phase flows in small-diameter channels and presents a comparison of calculation methods. For conditions of high reduced pressures $p_r = p/p_{cr} \approx 0.4 \div 0.6$, the results of own experimental studies of pressure drops and flow boiling heat transfer of freons in the region of low and high mass flow rates (G = 200–2000 kg/m^2 s) are presented. A description of the experimental stand is given, and a comparison of own experimental data with those obtained using the most reliable calculated relations is carried out.

Keywords: heat transfer; hydrodynamics; high reduced pressure; flow boiling

Citation: Belyaev, A.V.; Dedov, A.V.; Krapivin, I.I.; Varava, A.N.; Jiang, P.; Xu, R. Study of Pressure Drops and Heat Transfer of Nonequilibrial Two-Phase Flows. *Water* **2021**, *13*, 2275. https://doi.org/10.3390/w13162275

Academic Editors: Maksim Pakhomov and Pavel Lobanov

Received: 15 July 2021
Accepted: 17 August 2021
Published: 20 August 2021

1. Introduction

An important trend in the development of new energy conservation technologies is creating more miniature technical objects, an effort that requires extensive background knowledge of hydrodynamics and heat transfer in single-phase convection and flow boiling in mini-channels.

The opportunity to accurately predict the pressure drops and heat transfer and the selection of mini-channel geometry and working conditions are important factors for the design of the optimal settings of heat exchangers. In various fields of technology, one of the effective methods of heat transfer from heating surfaces is the boiling of liquid. It is necessary to experimentally confirm methods for calculating pressure drops and heat transfer.

1.1. Pressure Drops

The two-phase pressure drop in micro-channels is relatively high compared to conventional channels due to their very small size and relatively high mass flow rates, the latter being necessary to achieve acceptable heat transfer coefficients. Due to the high-pressure gradient, saturation temperature, and, consequently, thermophysical properties, there is a difference in pressure in mini-channels when the pressure in a certain axial position of the channel drops below the saturation pressure of the liquid, and the liquid temporarily overheats in this place. A pressure drop in a two-phase flow is a result of friction, acceleration, gravitation, and channel form change.

$$\Delta P_{TP} = \left[\left(\frac{dP}{dZ} \right)_{Fr} + \left(\frac{dP}{dZ} \right)_{Ac} + \left(\frac{dP}{dZ} \right)_{Gr} \right] Z + \Delta P_F \quad (1)$$

Gravitational pressure drop is usually neglected. If the flow is adiabatic, pressure drop due to flow acceleration is neglected too. Most techniques for calculating pressure drop relate to either a homogeneous or separated flow model.

1.1.1. Homogeneous Equilibrium Model

For a homogeneous equilibrium model, it is assumed that liquid and gas mix with each other, and the pressure drop of a two-phase flow can be calculated using the correlations for a single-phase flow. For this, the values averaged over the entire cross-section are taken as the calculated thermophysical properties, while there is heat transfer between the phases.

$$\left(\frac{dP}{dZ} \right)_{Fr} = \left(\frac{dP}{dZ} \right)_{TP} = \xi_{TP} \frac{(G)^2}{2\rho_{TP}} \frac{1}{D_h} \quad (2)$$

where ξ_{TP} is obtained by the Filonenko formula [1]:

$$\xi_{TP} = \frac{1}{(1.82 \log_{10}(Re_{TP}) - 1.64)^2} \quad (3)$$

$$Re_{TP} = \frac{GD_h}{\mu_{TP}} \quad (4)$$

μ_{TP} is calculated according to the method of Cicchitti et al. [2], which is the most popular method and researched for a wide range of the Reynolds numbers:

$$\mu_{TP} = x\mu_g + (1-x)\mu_l \quad (5)$$

As it said in Zubov et al. [3], in the limit of high velocities of the mixture at high reduced pressures, there is reason to draw an analogy between the homogeneous model and the continuum model for gas flow. The traditional homogeneous flow model for shear stress on the wall uses the formula:

$$\tau_{TP} = \frac{\xi_{TP}}{8} \rho_\beta w_{TP}^2 \quad (6)$$

where $\rho_\beta = \rho'' \beta + (1-\beta)\rho'$:

$$w_{TP} = \frac{G}{\rho'} \times \left[1 + x \frac{\rho' - \rho''}{\rho''} \right] \quad (7)$$

And then, pressure drops are calculated as:

$$\left(\frac{dP}{dZ} \right)_{Fr} = \frac{4\tau_{TP}}{D_h} \quad (8)$$

In two-phase flow, according to research by Venkatesan et al. [4], Cioncolini et al. [5], and Choi and Kim [6], the homogeneous flow model is applicable only to bubbly flow. Homogeneous flow conditions are fulfilled at high flow rates and mass flow rate steam content less than 0.1. At large values of the mass vapor quality $x > 0.1$, the homogeneous model, as a rule, is not applied. Under the conditions of subcooled flow, a calculation based on the homogeneous model demonstrates acceptable accuracy [7].

1.1.2. Separated Flow Model

Liquid and gas in the separated flow model move together, and it is taken as a fact that there is a clear phase boundary through which evaporation occurs. In the last decade, a number of studies have been published on pressure drop in micro-channels, based on the methodology of Lockhart and Martinelli [8], who proposed using the two-phase multiplier (9) to relate the pressure drop of a two-phase flow to the pressure drop of the liquid phase:

$$\Phi_l^2 = \left(\frac{dP}{dZ}\right)_{TP} \Big/ \left(\frac{dP}{dZ}\right)_l \tag{9}$$

An example of such research is the work of Hwang et al. [9], where the working fluid was R134a, and the hydraulic diameter varied from 0.244 to 0.792 mm. The study concluded that the pressure drop increased with an increase in the Reynolds number and was similar to the pressure drop for single-phase flow in a channel with a larger equivalent diameter. In addition, the pressure drop in a two-phase flow increases with a decrease in the inner diameter. The two-phase multiplier is calculated as:

$$\Phi_l^2 = 1 + \frac{C}{\chi} + \frac{1}{\chi^2} \tag{10}$$

where:

$$C = 227 Re_l^{0.452} Co^{-0.82} \chi^{-0.32} \tag{11}$$

To calculate the two-phase multiplier Φ_l in conventional channels, for example, the Friedel method is often used [10], in which pipes larger than 4 mm are examined. The two-phase multiplier is calculated as:

$$\Phi_l = E + \left(\frac{3.24FH}{Fr^{0.045} We^{0.035}}\right) \tag{12}$$

where:

$$E = (1-x)^2 + x^2 \frac{\rho_l}{\rho_g} \frac{f_g}{f_l} \tag{13}$$

$$F = x^{0.78}(1-x)^{0.224} \tag{14}$$

$$H = \left(1 - \frac{\mu_g}{\mu_l}\right)^{0.7} \left(\frac{\rho_l}{\rho_g}\right)^{0.91} \left(\frac{\mu_g}{\mu_l}\right)^{0.19} \tag{15}$$

Most formulas and methods, including the ones above, are suitable for relatively small amounts of fluid and a limited range of flow parameters and geometries. Thus, it is necessary to check the accuracy of the prediction models and select the best formula for predicting pressure drops in mini-channels.

1.2. Investigations of Heat Transfer in Mini-Channels

In the literature, there are many methods for determining the heat transfer coefficient when boiling a liquid flow in channels.

One of the best-known methods was by J. Chen [11], which was derived in a paper in which the boiling of a saturated water flow in a circular vertical micro-channel was investigated.

Lazarek and Black [12], when calculating the heat transfer coefficient, came to the conclusion that nucleate boiling was the main one that occurred during their tests, since the heat transfer coefficient depended on mass flow rate and heat flux density.

The experimental data in a study by Tran et al. [13] showed that when the value of vapor quality $x > 0.2$, the heat transfer coefficient does not depend on it. Here, heat transfer depended mainly on the mass velocity and not on the heat flux density. It was found that the border between the regions of the dominance of nucleate boiling and evaporation is rather abrupt and occurs at significantly smaller changes in saturation temperature than predicted.

Kenning-Cooper [14] noted that in the annular flow regime, the heat transfer coefficient is well described by J. Chen [11], but for slug flow, Chen's method gives deviations. In addition, nucleate boiling is sensitive to surface conditions as opposed to evaporative conditions.

Gungor and Winterton [15] developed a correlation that is versatile in its application and generally gives a more accurate fit to the data than the correlations proposed by the authors of the studies reviewed. The average deviation between the calculated and measured heat transfer coefficient was 21.4% for saturated boiling and 25.0% for supercooled boiling.

Shah [16] presented, in graphical form, a general correlation called CHAPT for estimating the saturated boiling heat transfer coefficients for subcritical heating of the flow in pipes.

Liu-Winterton [17] presented a comparative analysis of the previously obtained data by J. Chen [11], Gungor and Winterton [15], and Shah [16]. In their correlation, the authors introduced the Prandtl constraint as a parameter that affects the coefficient of influence of the convective component on the heat transfer coefficient.

Kandlikar [18] conducted a comparative analysis of the earlier studies concentrating on the data obtained by the different researchers. For the correlations, data from 24 experimental studies were obtained. For comparison, the correlations of J. Chen [11], Gungor and Winterton [15], and Shah [16] were considered.

Sun and Mishima [19] conducted a comparative analysis of 13 previously obtained correlations, forming a new database. The results showed that J. Chen's correlation [11] and its modifications were not very well suited for mini-channels and that Lazarek and Black's correlation [12] was the most suitable.

Table 1 summarizes the most famous works. Obviously, the data in most of the experimental works now available in the literature were obtained for low and moderate reduced pressures [20]. In addition, the authors proposed the calculation methods, which have an empirical nature and are more suitable for describing experiments close to those described by the authors. In the field of high reduced pressures, the analysis of the literature shows a lack of researches.

Table 1. List of works indicating calculation methods.

Author	Liquids	Formula
Tran et al.	R12, R113	$\alpha_{TP} = 840\left(Bo^2 We_{l\ only}\right)^{0.3}\left(\rho_g/\rho_l\right)^{0.4}$
Lazarek and Black	R113	$\alpha_{TP} = 30Re_l^{0.857}Bo^{0.714}\frac{\lambda_l}{D_h}$
Shah	R11, R12, R22	$\alpha_{TP} = \psi\alpha_{SP}$
Kenning-Cooper	Water, freons	$\alpha_{TP} = \left(1 + 1.8\chi_{tt}^{-0.87}\right)\alpha_{SP}$
Kandlikar	Water, R11, R12, R114, NO_2	$\frac{\alpha_{TP}}{\alpha_{SP}} = \begin{cases} 1.136Co^{-0.9}(25Fr_l)^c + 667.2Bo^{0.7}F_l\ (1) \\ 0.0683Co^{-0.2}(25Fr_l)^c + 1058Bo^{0.7}F_l\ (2) \\ (1): Co< 0.65; (2): Co >0.65 \end{cases}$
Sun and Mishima	Water, freons	$\alpha_{TP} = \frac{6Re_{lo}^{1.05}Bo^{0.54}}{We_{lo}^{0.191}\left(\frac{\rho_l}{\rho_g}\right)^{0.142}}\frac{\lambda_l}{D_h}$
J. Chen	Water	$\alpha_{TP} = S\alpha_{NB} + F\alpha_{SP}$
Gungor and Winterton	Water, R22, R113, R114, R11, R12	$\alpha_{TP} = S\alpha_{NB} + F\alpha_{SP}$
Liu-Winterton	Water, R113, R114, R11, R12, R22	$\alpha_{TP} = \sqrt{(S\alpha_{NB})^2 + (F\alpha_{CB})^2}$

The size of the channel significantly affects the character of vaporization during flow boiling. In the region of high reduced pressures, based on the analysis performed in [21], it can be assumed that, in mini-channels, the flow regimes become identical to those seen in conventional channels. In this case, the relationships for the normal channels may be used to calculate the pressure drop and heat transfer. Based on this assumption, a method for calculating heat transfer for subcooled flow boiling in mini-channels was tested in [7].

The heat flux density was calculated as follows:

$$q = q_{boil} + q_{con} \tag{16}$$

It is assumed that convective heat transfer acts in the same way as in a single-phase turbulent flow:

$$q_{con} = \alpha_{con}(T_{wall} - T_{fluid}) \tag{17}$$

where α_{con} is calculated using the Petukhov formula with employees in the form [22], adjusted for the difference between the wall and liquid temperatures:

$$Nu = \frac{(\xi/8)(Re - 1000)Pr}{1 + 12.7(\xi/8)^{1/2}(Pr^{2/3} - 1)} \left(\frac{Pr_l}{Pr_{wall}}\right)^{0.25} \tag{18}$$

To calculate q_{boil} in conditions of saturated flow boiling in relation (16), it is advisable to use the equation proposed by V.V. Yagov [23]:

$$q_{boil} = 3.43 \times 10^{-4} \frac{\lambda^2 \Delta T_s{}^3}{\nu\sigma T_s} \left(1 + \frac{r\Delta T}{2R_i T_s^2}\right)\left(1 + \sqrt{1 + 800B} + 400B\right) \tag{19}$$

where $B = \frac{r\left(\rho_g \frac{\mu_l}{\rho_l}\right)^{3/2}}{\sigma(\lambda T_s)^{1/2}}$ and $\Delta T_s = T_{wall} - T_s$ (all properties are determined at saturation temperature T_s). Modified version of Equation (19) for q_{boil} for subcooled flow boiling presented in the paper [24].

2. Experimental Setup Description

The scheme of the experimental setup is shown in Figure 1. The hydraulic circuit allows maintaining stable flow parameters at pressures up to 2.7 MPa and temperatures up to 150 °C. A multistage centrifugal pump was used for the creation of working fluid circulation (location 6 in Figure 1). The mass flow rate was measured with a high-precision coriolis flowmeter (location 7).

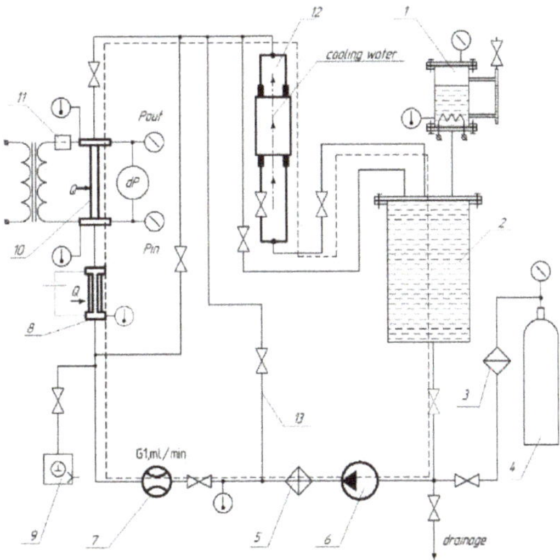

Figure 1. Experimental setup: (1) thermocompressor, (2) tank, (3) and (5) filters, (4) balloon with refrigerant, (6) multistage centrifugal pump, (7) coriolis flowmeter, (8) pre-heater, (9) roughing-down pump, (10) test section, (11) current transducer, (12) recuperative heat exchanger, (13) bypass line.

The working fluids in this study was R125, which has a critical temperature of 66.023 °C and a critical pressure of 3.6177 MPa. Heat capacity, heat of vaporization and critical pressure of R125 are much lower than that of water, which is very convenient for achieving the desired parameters. The working fluid was cooled by water in a recuperative heat exchanger (location 12). High reduced pressure in the circuit was created by using a thermocompressor (location 1). A pressure sensor with a measurement accuracy of 0.2% was used for measuring pressure and pressure drops across the inlet and outlet of the test section. Chromel-Copel cable thermocouples with a cable diameter of 0.7 mm measured the inlet and outlet temperatures.

The test section was heated with alternating current. The electrical current strength was measured using an LA 55-P current transducer. The measurement error of the electric power was 1%.

The test section is shown in Figure 2. Vertical stainless-steel tubes with heated lengths of 51 mm each and internal diameters of 1 mm and 1.1 mm were used as mini-channels. The tube was electrically insulated and hydraulically sealed through PTFE (polytetrafluoroethylene) seals. Electrodes were soldered to the tube with tin.

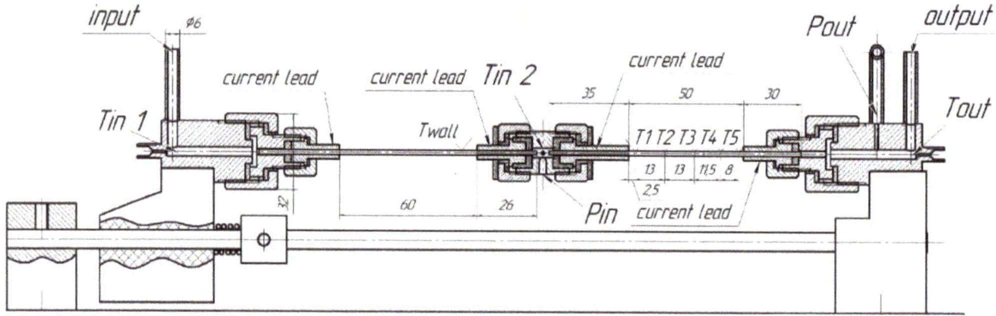

Figure 2. Design of the test section.

The design of the test section had temperature-compensation. The platform with the inlet collector was mounted on two vertical metal rods on which it could slide. In this way, the inlet collector had a vertical degree of freedom. The platform of the inlet collector was held by a spring along the rods towards the tube to avoid vibration and ensure the stability of the test tube.

Five Chromel-Copel thermocouples were used to take the measure values of the wall temperatures. On five cross-sections (T1–T5, see Table 2) of the working area of the tube on opposite sides of the tube diameter, the wires (diameter 0.2 mm) were welded using lasers. This mounting method for the thermocouples created low thermal inertia for the sensors and allowed the measurement of the average temperature of the wall along its perimeter. The inner wall temperatures were calculated using a correction for the wall conductivity.

Table 2. Coordinates of the cross-sections.

Diameter (mm)	T1	T2	T3	T4	T5
1.0	-	-	15	30	45
1.1	2.5	15.5	28.5	40	48

3. Pressure Drop

In this study, experimental data on pressure drop for a range of mass flow rates $G = 200–2000\ \mathrm{kg/(m^2\ s)}$ were obtained at two channels with diameters 1.0 and 1.1 mm. The data were obtained for a wide range of heat flux density, which made it possible to obtain bubble and film flow regimes.

Figures 3 and 4 show the primary pressure drop data. For most of the obtained characteristics, the $\Delta p(q)$ regions of various flow regimes were observed, such as: convective heat transfer, when the pressure drop remained almost unchanged; nucleate boiling with an intense increase in pressure drop; film boiling regime, when the growth of pressure drops stopped.

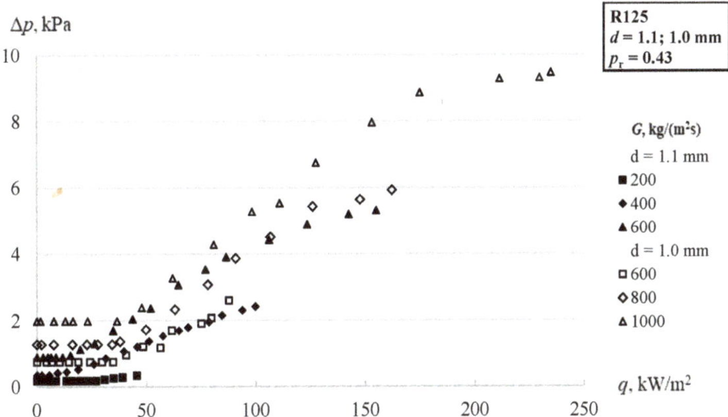

Figure 3. Pressure drop versus heat flux density at various values of the mass flow rates.

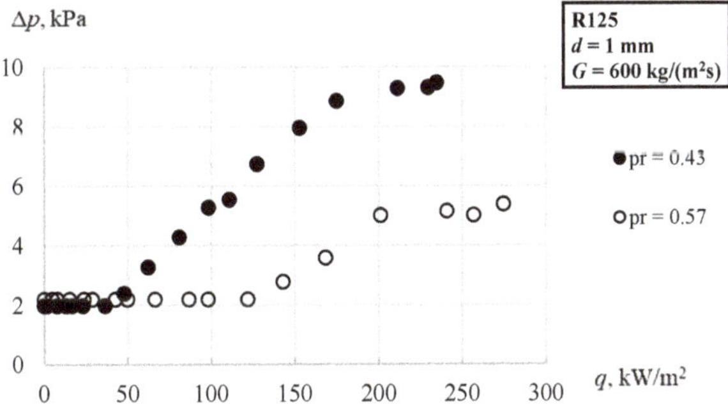

Figure 4. Pressure drop versus heat flux density at two values of reduced pressure.

With an increase in the diameter, the pressure drops decreased for the same values of mass flow rate (see Figure 3), which is quite natural. Analysis of the effect of reduced pressure on the pressure drops at the same values of G = 600 kg/(m^2 s) (see Figure 4) allowed us to draw the following conclusion: the flow regime changed with an increase in the reduced pressure: the region with non-increasing pressure drops due to the heat load increases from 50 to 125 kW/m^2.

To obtain pressure drops by calculation, the earlier described methods were used. In most of the experiments, the calculation method by [10] had too significant deviation with increasing heat flux, as can be seen in Figure 5, probably due to the quadratic dependence of the two-phase multiplier Φ_l on the vapor quality x. In addition, this method has been developed for channels with diameters greater than 4 mm. As a result, it was concluded that it was not suitable for generalizing the obtained data on mini-channels, even under conditions of high reduced pressures.

The analysis of the experimental data showed that the reduced pressure mostly affected the correspondence of the calculated values to the experimental data. For the investigated range of mass flow rates G = 200–2000 kg/(m^2 s) and values of vapor quality (up to $x \approx 0.4$) for reduced pressure p_r = 0.43, the best agreement with the experimental data was observed for the method of [9], which was based on a split flow model. An example of calculations for pressure p_r = 0.43 is shown in Figure 6.

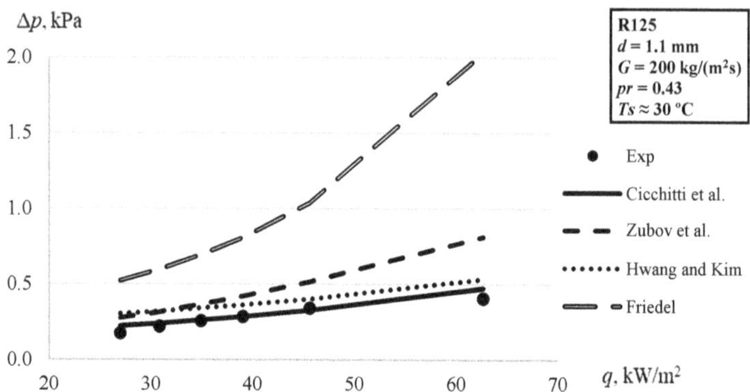

Figure 5. Example of the experimental data with calculated data versus heat flux.

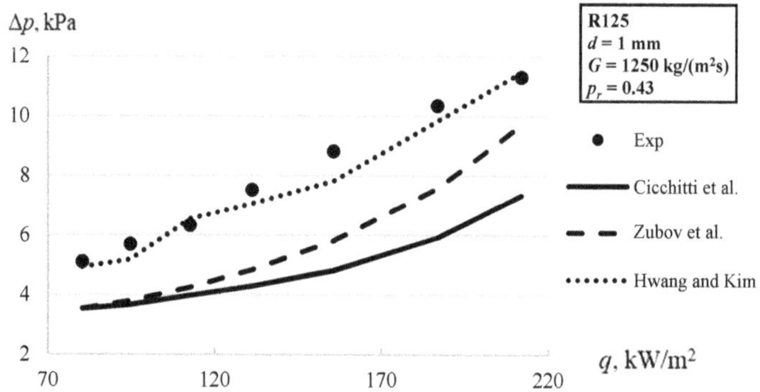

Figure 6. Pressure drop versus heat flux for experimental and calculated data at $G = 1250\ \text{kg/m}^2\ \text{s}$ and $p_r = 0.43$.

For the data obtained at reduced pressure $p_r = 0.57$, the calculation using the homogeneous models of [2] and [3] was in better agreement with the experiment than the calculation using the split flow model. Figure 7 shows an example of a calculation for $p_r = 0.57$ and average mass flow rate $G = 750\ \text{kg/(m}^2\ \text{s)}$.

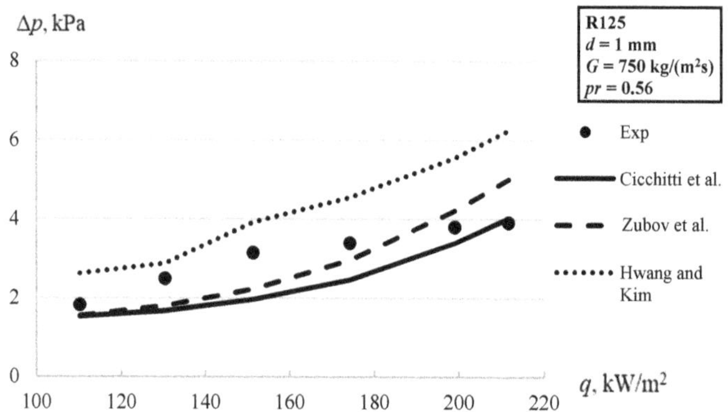

Figure 7. Pressure drop versus heat flux for experimental and calculated data at $G = 750\ \text{kg/m}^2\ \text{s}$ and $p_r = 0.56$.

The following Table 3 presents the generalization of all experimental pressure drop data summarized by the three considered calculation methods. The data are divided into two groups according to the values of the reduced pressure.

Table 3. Comparison of obtained pressure drop databases with predictions of selected correlation.

p_r	Deviation	0–10%	0–20%	0–30%
0.43	Cicchitti et al. [2]	14%	23%	37%
	Zubov et al. [3]	13%	33%	58%
	Hwang and Kim [9]	29%	58%	84%
0.57	Cicchitti et al. [2]	13%	42%	71%
	Zubov et al. [3]	17%	46%	88%
	Hwang and Kim [9]	0%	4%	13%

From the analysis of the generalization of the obtained experimental data on pressure drop, it can be concluded that there was a significant effect of reduced pressure on the agreement of the calculated values obtained using the methods of [2,3,9] with the experimental data. As can be seen from Table 3, the homogeneous model was more suited to high reduced pressures, and the split flow model showed a good result at lower reduced pressure. This is probably due to a change in the structure of flow boiling with an increase in pressure as a result of a decrease in the diameter of the vapor bubble.

4. Flow Boiling Heat Transfer

Primary data on heat flux based on wall overheating relative to the saturation temperature for different mass flow rates are shown in Figure 8. At $G \leq 1750$ kg/(m^2 s), the contribution of convective heat transfer to total heat transfer was insignificant, and the boiling curves lay close to each other with a temperature deviation of about 1 °C. With increasing G, the contribution of convective heat transfer to total heat transfer became significant, which is quite natural, and the boiling curve for G = 2000 kg/(m^2 s) was significantly higher than for other points. Thus, nucleate boiling was obviously the main mechanism of heat transfer at the given mass flow rates.

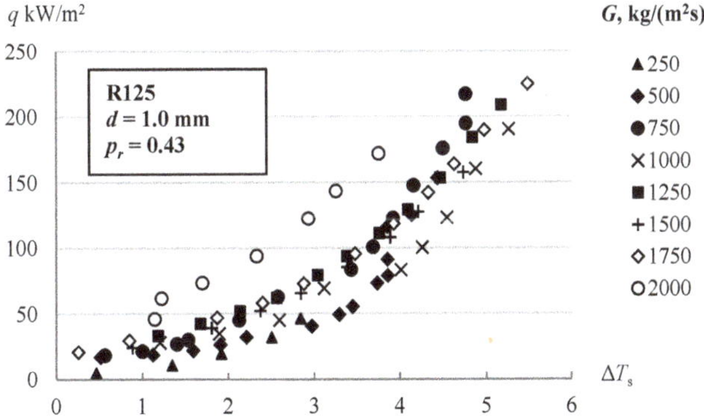

Figure 8. Experimental data for heat flux versus wall overheating at various mass flow rates.

The dependence of the heat transfer coefficient on the heat flux density for one mode in comparison with the calculation results given by the Petukhov Formula (18) for convective heat transfer is shown in Figure 9. It was possible to obtain a small area of convective heat transfer data points due to the impossibility of making the temperature at the entrance of the test section and, consequently, at subcooling below the room temperature. However, it can be seen from Figure 9 that the calculated values coincided with the experimental data in the region of convective heat transfer, which makes it possible to verify the experimental data.

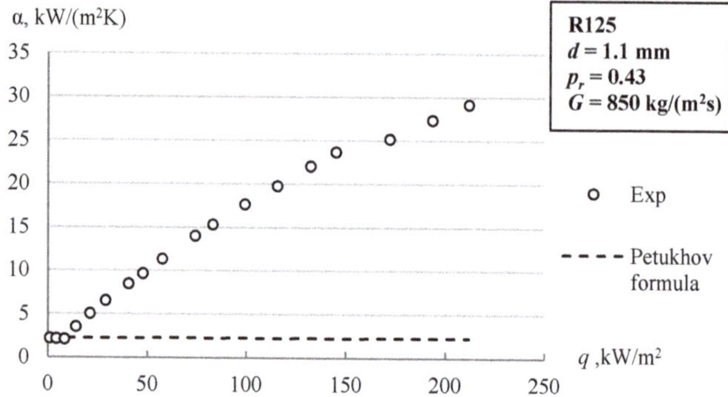

Figure 9. Heat transfer coefficient versus heat flux.

Comparison of the data obtained from calculations using Formulas (16)–(19) with the primary experimental data for a low mass flow rate and saturated liquid is shown in Figure 10. A comparison example of the calculation with the experimental data from [7] and [25], corresponding to moderate subcooling and high mass flow rate, is shown in Figure 11.

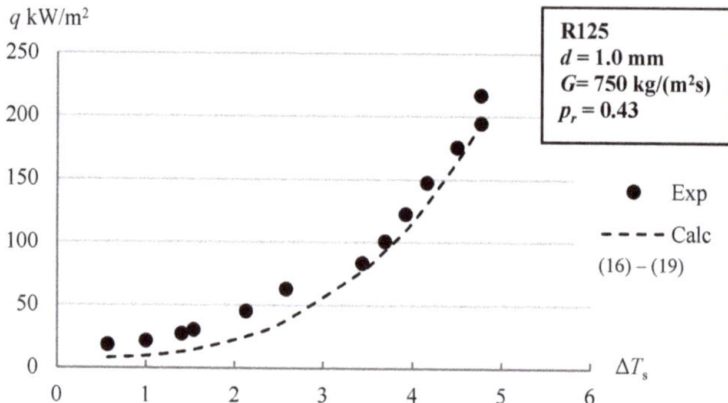

Figure 10. Comparison of the calculated data for heat transfer with the experimental data for saturated liquid ($x_{\text{local}} \approx 0 \div -0.4$).

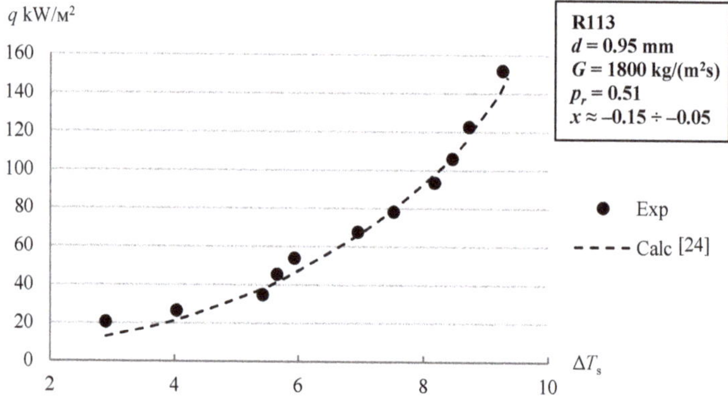

Figure 11. Comparison of the experimental data for subcooled liquid ([7,25], $x_{\text{local}} \approx -0.15$ to -0.05) with the calculated heat transfer data.

The graphs show that the calculated values were in good agreement with the experimental data. One of the features of the subcooled flow boiling, as can be seen from the primary data obtained, was a higher wall overheating (see Figure 11) compared with the saturated boiling (see Figure 10).

Figure 12 shows the data obtained in the current study for the most requested range of low and moderate mass flow rates G = 200–1000 kg/(m^2 s). Generalization was performed using Formulas (16–19). The calculation results were in good agreement with the experimental data for $x > 0$, and the mean absolute error is 16%.

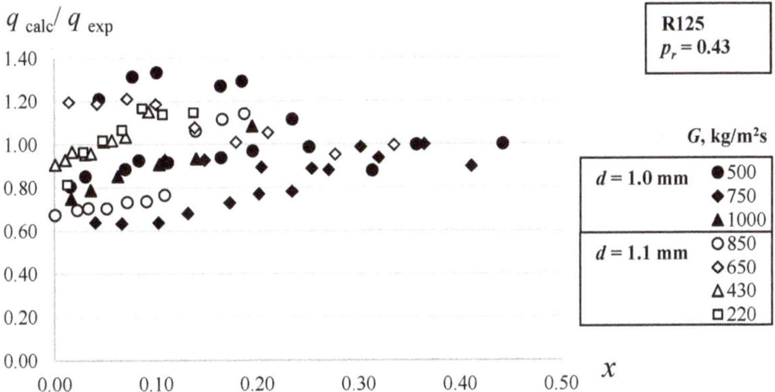

Figure 12. Comparison of the experimental data with the calculated data obtained using Formulas (16)–(19).

5. Conclusions

This paper has presented an experimental setup along with the results of an investigation of the heat transfer and pressure drop during flow boiling of R125 in two vertical channels with diameters 1.0 and 1.1 mm and lengths 51 mm each under different combinations of high reduced pressure, mass flow rate, and heat flux. These parameters were varied within the following ranges: reduced pressure $p_r \approx 0.4$–0.6, mass flow rate G = 200–2000 kg/(m^2 s), and heat flux q from boiling onset to crisis.

The most popular methods in the literature for calculating pressure drop and heat transfer during flow boiling in mini-channels have been analyzed. The analysis shows a practical lack of researches with experiments at high reduced pressures.

Generalization of own data on pressure drop and heat transfer has been performed. Based on the literature review, calculation methods of [2,3,9] were chosen for the generalization of the pressure drop data. From the analysis of the generalization, it was concluded that the homogeneous model was more suited for high reduced pressures, and the split flow model showed a good result at lower reduced pressures. Thus, at a higher reduced pressure, the flow regime was more similar to the homogeneous model, whereas at a lower pressure, the flow structure was more similar to the split flow model. No such effect of the mass flow rate on flow structure was observed.

The methods for calculating pressure drop during flow boiling require further elaboration. It is necessary to establish the limits of applicability of various types of models for calculating pressure drop, depending on the reduced pressure and the degree of saturation of fluid flow.

To generalize the data on heat transfer, the previously approved method [7] was used with the division of the calculation of heat flux into convection and nucleate boiling heat flux. The presented calculation method, which are based on Formulas (16)–(19), satisfied the obtained experimental results of heat transfer with 16% of mean absolute error. This method can be applied in the most requested range of mass flow rates G = 200–1000 kg/m^2 s and $x > 0$.

Author Contributions: Data curation, A.V.B.; formal analysis, A.V.B.; investigation, A.V.B. and R.X.; methodology, A.V.B., A.V.D., A.N.V. and P.J.; writing—original draft preparation, A.V.B. and I.I.K.; conceptualization, A.V.D., P.J. and R.X.; project administration, A.V.D.; validation, A.V.D. and P.J.; writing—review and editing, A.V.D. and R.X.; software, I.I.K.; formal analysis, A.N.V. and P.J.; supervision, A.N.V. All authors have read and agreed to the published version of the manuscript.

Funding: This work was supported by RSF Grant 19-19-00410.

Conflicts of Interest: The authors declare no conflict of interest.

Nomenclature

d	diameter, m
G	mass flow rate, kg/(m^2 s)
p	pressure, Pa
T	temperature, K
x	vapor quality
c_p	specific heat, J/(kg·K)
r	latent heat of evaporation, J/kg
w	velocity, m/c^2
Greek symbols	
α	heat transfer coefficient, W/m^2·K
σ	surface tension, N/m
λ	thermal conductivity, W/(m·K)
ξ	hydraulic friction factor
ρ	density, kg/m^3
β	volume vapor quality
μ	viscosity, N·s/m^2
χ	Martinelli parameter
We	Weber number $We = \frac{G^2 D_h}{\rho\sigma}$
E; F; H	Friedel parameters
Fr	Froude number $Fr = \frac{w^2}{g D_h}$
Φ	two-phase multiplier
Co	confinement number $Co = \left(\frac{\sigma}{g(\rho_l - \rho_g)}\right)^{0.5} D_h^{-1}$
Subscripts	
l	liquid
g	gas
boil	boiling
con	convective
calc	calculated
exp	experimental
sub	subcooled
cr	critical
in	inlet
s	saturated
r	reduced
Fr	friction
TP	two-phased
SP	single phase
CB	convective boiling

References

1. Filonenko, G.I. Gidravlicheskoe soprotivlenie truboprovodov. *Teploenergetika* **1954**, *4*, 40.
2. Cicchitti, A.; Lombardi, C.; Silvestri, M.; Soldaini, G.; Zavatarelli, R. Two-phase cooling experiments-pressure drop, heat transfer and burnout measurements. *Energ. Nucl.* **1960**, *7*, 407–425.
3. Zubov, N.O.; Kaban'Kov, O.N.; Yagov, V.V.; Sukomel, L.A. Prediction of friction pressure drop for low pressure two-phase flows on the basis of approximate analytical models. *Therm. Eng.* **2017**, *64*, 898–911. [CrossRef]

4. Venkatesan, M.; Das, S.K.; Balakrishnan, A. Effect of diameter on two-phase pressure drop in narrow tubes. *Exp. Therm. Fluid Sci.* **2011**, *35*, 531–541. [CrossRef]
5. Cioncolini, A.; Thome, J.R.; Lombardi, C. Unified macro-to-microscale method to predict two-phase frictional pressure drops of annular flows. *Int. J. Multiph. Flow* **2009**, *35*, 1138–1148. [CrossRef]
6. Choi, C.; Kim, M. Flow pattern based correlations of two-phase pressure drop in rectangular microchannels. *Int. J. Heat Fluid Flow* **2011**, *32*, 1199–1207. [CrossRef]
7. Belyaev, A.; Varava, A.; Dedov, A.; Komov, A. An experimental study of flow boiling in minichannels at high reduced pressure. *Int. J. Heat Mass Transf.* **2017**, *110*, 360–373. [CrossRef]
8. Lockhart, R.W.; Martinelli, R.C. Proposed correlation of data for isothermal two-phase, twocomponent flow in pipes. *Chem. Eng. Prog.* **1949**, *45*, 39–48.
9. Hwang, Y.W.; Kim, M.S. The pressure drop in microtubes and the correlation development. *Int. J. Heat Mass Transf.* **2006**, *49*, 1804–1812. [CrossRef]
10. Friedel, L. Improved friction pressure drop correlations for horizontal and vertical two-phase pipe flow. In Proceedings of the European Two-Phase Flow Group Meeting, Ispra, Italy, 5–8 June 1979; Paper E2.
11. Chen, J.C. Correlation for Boiling Heat Transfer to Saturated Fluids in Convective Flow. *Ind. Eng. Chem. Process. Des. Dev.* **1966**, *5*, 322–329. [CrossRef]
12. Lazarek, G.; Black, S. Evaporative heat transfer, pressure drop and critical heat flux in a small vertical tube with R-113. *Int. J. Heat Mass Transf.* **1982**, *25*, 945–960. [CrossRef]
13. Tran, T.; Wambsganss, M.; France, D. Small circular- and rectangular-channel boiling with two refrigerants. *Int. J. Multiph. Flow* **1996**, *22*, 485–498. [CrossRef]
14. Kenning, D.B.R.; Cooper, M.G. Saturated flow boiling of water in vertical tubes. *Int. J. Heat Mass Transf.* **1989**, *32*, 445–458. [CrossRef]
15. Gungor, K.; Winterton, R. A general correlation for flow boiling in tubes and annuli. *Int. J. Heat Mass Transf.* **1986**, *29*, 351–358. [CrossRef]
16. Shah, M.M. Chart correlation for saturated boiling heat transfer: Equations and further study. *ASHRAE Trans.* **1982**, *88*, 185–196.
17. Liu, Z.; Winterton, R. A general correlation for saturated and subcooled flow boiling in tubes and annuli, based on a nucleate pool boiling equation. *Int. J. Heat Mass Transf.* **1991**, *34*, 2759–2766. [CrossRef]
18. Kandlikar, S.G. A General Correlation for Saturated Two-Phase Flow Boiling Heat Transfer Inside Horizontal and Vertical Tubes. *J. Heat Transf.* **1990**, *112*, 219–228. [CrossRef]
19. Sun, L.; Mishima, K. An evaluation of prediction methods for saturated flow boiling heat transfer in mini-channels. *Int. J. Heat Mass Transf.* **2009**, *52*, 5323–5329. [CrossRef]
20. Kim, S.-M.; Mudawar, I. Review of databases and predictive methods for heat transfer in condensing and boiling mini/micro-channel flows. *Int. J. Heat Mass Transf.* **2014**, *77*, 627–652. [CrossRef]
21. Yagov, V.V.; Minko, M.V. Teploobmen v dvukhfaznom potoke pri vysokikh privedennykh davleniyakh. *Teploehnergetika* **2011**, *4*, 13–23.
22. Gnielinski, V. New equations for heat and mass transfer in turbulent pipe and channel flow. *Int. J. Chem. Eng.* **1976**, *16*, 359–368.
23. Yagov, V.V. Teploobmen pri razvitom puzyr'kovom kipenii. *Teploenergetika* **1988**, *2*, 4–9.
24. Dedov, A.V.; Komov, A.T.; Varava, A.N.; Yagov, V.V. Hydrodynamics and heat transfer in swirl flow under conditions of one-side heating. Part 2: Boiling heat transfer. *Crit. Heat Fluxes Int. J. Heat Mass Transf.* **2010**, *53*, 4966–4975.
25. Belyaev, A.; Varava, A.; Dedov, A.; Komov, A. Critical heat flux at flow boiling of refrigerants in minichannels at high reduced pressure. *Int. J. Heat Mass Transf.* **2018**, *122*, 732–739. [CrossRef]

Article

The Flow Pattern Transition and Water Holdup of Gas–Liquid Flow in the Horizontal and Vertical Sections of a Continuous Transportation Pipe

Guishan Ren [1], Dangke Ge [1], Peng Li [2,*], Xuemei Chen [1], Xuhui Zhang [2,3], Xiaobing Lu [2,3], Kai Sun [1], Rui Fang [1], Lifei Mi [1] and Feng Su [1]

1 Oil Production Technology Institute, Dagang Oilfield, Tianjin 300280, China; rengshan@petrochina.com.cn (G.R.); gedke@petrochina.com.cn (D.G.); chenxmei@petrochina.com.cn (X.C.); sunkai18@petrochina.com.cn (K.S.); fangr@petrochina.com.cn (R.F.); milfei@petrochina.com.cn (L.M.); sufeng@petrochina.com.cn (F.S.)
2 Institute of Mechanics, Chinese Academy of Sciences, Beijing 100190, China; zhangxuhui@imech.ac.cn (X.Z.); xblu@imech.ac.cn (X.L.)
3 School of Engineering Science, University of Chinese Academy of Sciences, Beijing 100049, China
* Correspondence: lipeng@imech.ac.cn

Citation: Ren, G.; Ge, D.; Li, P.; Chen, X.; Zhang, X.; Lu, X.; Sun, K.; Fang, R.; Mi, L.; Su, F. The Flow Pattern Transition and Water Holdup of Gas–Liquid Flow in the Horizontal and Vertical Sections of a Continuous Transportation Pipe. *Water* **2021**, *13*, 2077. https://doi.org/10.3390/w13152077

Academic Editors: Maksim Pakhomov and Pavel Lobanov

Received: 19 June 2021
Accepted: 26 July 2021
Published: 30 July 2021

Abstract: A series of experiments were conducted to investigate the flow pattern transitions and water holdup during oil–water–gas three-phase flow considering both a horizontal section and a vertical section of a transportation pipe simultaneously. The flowing media were white mineral oil, distilled water, and air. Dimensionless numbers controlling the multiphase flow were deduced to understand the scaling law of the flow process. The oil–water–gas three-phase flow was simplified as the two-phase flow of a gas and liquid mixture. Based on the experimental data, flow pattern maps were constructed in terms of the Reynolds number and the ratio of the superficial velocity of the gas to that of the liquid mixture for different Froude numbers. The original contributions of this work are that the relationship between the transient water holdup and the changes of the flow patterns in a transportation pipe with horizontal and vertical sections is established, providing a basis for judging the flow patterns in pipes in engineering practice. A dimensionless power-law correlation for the water holdup in the vertical section is presented based on the experimental data. The correlation can provide theoretical support for the design of oil and gas transport pipelines in industrial applications.

Keywords: oil–water–gas flow; flow pattern; water holdup; dimensionless analysis

1. Introduction

The pipe transportation of oil–water–gas three-phase systems is a crucial process in oil and natural gas production and provides vital information for interpreting the production stages. A deep understanding of the flow characteristics, such as the flow patterns and water holdup (the volume fraction of water in a pipe section), of the three-phase flow is beneficial to the proper design and operation of pipelines [1]. The different flow patterns directly determine the different flow characteristics of multiphase flows. This is a notable feature of multiphase flows in pipes, and it is an essential topic in multiphase flow research. The change of the flow pattern has a vital impact on the pressure drop (reflected in the energy consumption of transportation), spatial phase distribution, and safety of pipeline transportation [2]. For example, in the churn flow of gas–liquid flow, the bubbles have different sizes and shapes, and the liquid film attached to the pipe wall becomes an up–down vibrating flow, which affects the stability of the pipeline flow. To estimate the frictional pressure gradient accurately in a transparent vertical pipe, Xu et al. [3] studied the actual flow pattern under specific flow conditions. The flow patterns can be used to deduce the concentration distribution of each phase. Jones and Zuber [4] demonstrated that the probability density function (PDF) of the fluctuations in the volume fraction could be used as a statistical analysis tool for flow pattern identification. In short, it is crucial to predict the flow patterns and flow pattern transitions in oil pipeline transportation.

The calculation of water holdup is useful for predicting the quantity of oil in a petroleum pipeline, and water holdup is an important parameter for the classification of flow patterns. For example, Hasan and Kabir [5] proposed a semi-mechanistic method based on the flow pattern map to predict the in situ oil volume fraction and pressure drop. This model can interpret production logs to predict oil/water production rates. Liu et al. [6] developed a new annular flow model for calculating low water holdup in a horizontal pipe. Du et al. [7] experimentally investigate vertical upward oil–water two-phase flow in a 20 mm inner diameter pipe. The water holdup was measured using a vertical multiple electrode array conductance sensor, and five observed oil–water two-phase flow patterns were defined using mini-conductance probes.

Many researchers have studied the simultaneous oil–water–gas flow in a horizontal or vertical pipe. Oddie et al. [8] conducted steady-state and transient experiments of oil–water–gas multiphase flows in a large-diameter inclined pipe (11-m length, 15-cm diameter). The pipe inclination was varied from 0° (vertical) to 92°, and the flow rates of each phase were varied over wide ranges. A nuclear densitometer was used to measure the steady-state holdup values, and 10 electrical conductivities were used to provide transient and steady-state holdup profiles. The relationship between the measured holdup and flow rates, flow pattern, and pipe inclination was discussed. Spedding et al. [9] reported flow regimes for horizontal co-current oil-water-air three-phase flow for two different diameters. Combinations of dimensionless numbers for each phase were used as the mapping parameters. Two horizontal experimental three-phase facilities were used, and the flow patterns were identified using a combination of visual/video observations. Descamps et al. [10] performed laboratory experiments on the oil–water–air flow through a vertical pipe to study the gas-lifting technique for oil–water flows. The pressure gradient of the three-phase flow was always smaller than that of oil–water flow due to the air injection, except at the point of phase inversion. Air injection did not affect the concentration of oil and water at the phase inversion point. Hanafizadeh et al. [11] conducted experiments of air–water–oil three-phase flow patterns in an inclined pipe and investigated the effect of the liquid volume fraction and inclination angle on the flow patterns. The results showed that by increasing the oil cut for different inclination angles, the bubbly region was extended, and the plug region became smaller.

Numerous studies have applied computational fluid dynamics (CFD) approaches to simulate the hydrodynamics of multiphase pipe flow [12,13]. Ghorai et al. [14] developed a mathematical model to predict the holdup and pressure gradient for the water–oil–gas stratified flow in a horizontal pipe. The variations of the water holdup and pressure gradient for different situations were studied. However, the analysis was based on horizontal stratified flow only. Friedemann et al. [15] conducted a series of simulations on the gas–liquid slug flow in a horizontal concentric annulus using OpenFOAM and the built-in volume of fluid (VOF)-type solver interFoam. The simulation data were analyzed in terms of the pressure gradient and holdup profile, and they were compared to experimental data. The corresponding relationship between the flow pattern and water holdup was established. Leporini et al. [16] presented a new sand transport model implemented in one-dimensional dynamic multiphase code to deal with the liquid–solid flow as well as gas–liquid–solid flow. The numerical results demonstrated a good agreement with experimental data.

However, few studies on the comparison of the horizontal and vertical flow in a continuous transportation pipe have been reported [15,17]. In an on-site oil pipeline layout, pipelines are horizontal, vertical, and even inclined. Under the same flow parameters, pipelines in different directions may show different flow pattern characteristics, thereby creating hidden dangers for pipeline transportation stability. It is necessary to study the different flow behaviors in horizontal and vertical pipes under the same flow parameters.

The oil–water–gas flow is highly complex and related to the pipe geometry (e.g., inner pipe diameter and pipe angle), fluid properties (e.g., viscosity, density, and surface tension), and boundary conditions (e.g., superficial input velocities) [18]. Previous studies mainly focused on the effect of a single parameter, but the coupled effect of the controlling parameters on the flow is not well understood. Hence, to understand the fundamental mechanisms of oil–water–gas flows, the controlling dimensionless parameters were derived by dimensional analysis first. In a three-phase flow, it is difficult to distinguish the boundary between oil and water, especially at high flow rates. Thus, the oil–water–gas three-phase flow can be simplified as the two-phase flow of a gas and liquid mixture considering, as the densities of oil and water are much higher than that of gas and the oil and water velocities are sufficiently high to obtain a mixture [11].

Thus, the objective of this work was to investigate the flow patterns and water holdup for a simplified gas–liquid flow in horizontal and vertical sections through a comparative study of the flow behaviors in a pipe loop considering different superficial input velocities. In particular, a series of experiments were conducted to investigate the flow pattern transition and the water holdup, considering both the horizontal and vertical sections of a transportation pipe. In addition, the relationship between the transient water holdup and the change of the flow pattern in a transportation pipe with horizontal and vertical sections was established, which provides a basis for judging the flow pattern in a pipe in engineering practice. A dimensionless power-law correlation for the water holdup in the vertical section is presented based on the experimental data.

The current paper is structured as follows. The experimental setup is described in Section 2. In Section 3, the dimensionless numbers are derived based on the physical analysis and the proper choice of the units for the problem. In Section 4, the relationship between the transient water holdup and the change of the flow pattern in a transportation pipe with horizontal and vertical sections is established, and a dimensionless power-law correlation for the water holdup in the vertical section is presented. Finally, Section 5 presents the conclusions of this study.

2. Description of Experiments

Figure 1 shows the schematic diagram of the oil–water–gas three-phase flow loop, which consisted of a power system, a metering system, and a mixing line. All the experiments were conducted using white mineral oil, distilled water, and air. Yellow dye was added to the oil to differentiate it from water visually. In the experiments, the temperature of the environment and the experimental section were measured by temperature sensors (Rosemount, 248 type). During the experiment, which lasted 10 h per day, the temperature of the experimental section varied from 19 °C to 22 °C. There was a long flow pattern development section between the pumps used for the fluid (water and oil) and the experimental section; so, the pumps had little effect on the fluid temperature. The changes in the physical properties of the fluid were not significant. For the sake of simplicity, the physical properties of each phase in this study were determined at atmospheric pressure and room temperature 20 °C. The physical properties of the fluids tested are presented in Table 1. The values were selected based on the experimental results of Wang [19].

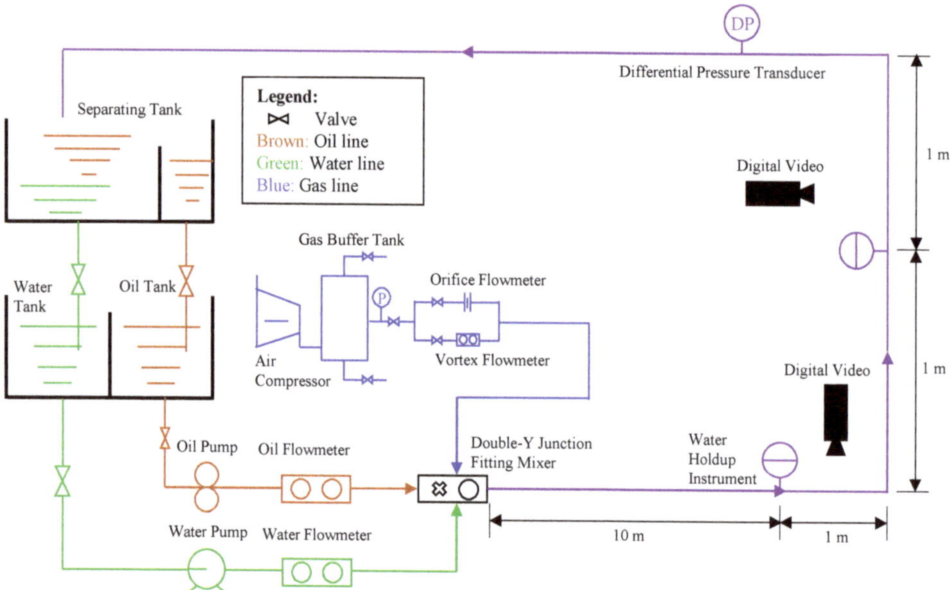

Figure 1. Schematic drawing of the flow loop for the oil–water–gas three-phase flow used in this study.

Table 1. Properties of water, oil, and gas phases measured at atmospheric pressure and 20 °C [20].

	Density, ρ (kg/m^3)	Viscosity, μ (mPa s)	Surface Tension with the Gas Phase, σ (mN m)
White mineral oil	841	28	30
Distilled water	998.2	1	72
Air	1.2	0.018	-

According to Wang's experimental measurements [19], the relationship between the density and temperature of white mineral oil is as follows in Equation (1):

$$\rho_o = 854.878 - 0.703t \tag{1}$$

where ρ_0 is the oil density, and t is the temperature (°C).

The relationship between the viscosity and temperature of white mineral oil is as follows in Equation (2):

$$\mu_o = 6.075 + 67.192exp(-t/18.2) \tag{2}$$

where μ_o is the oil viscosity.

The surface tension of white mineral oil is expressed as follows in Equation (3):

$$\sigma_o = 31.77 - 0.078t \tag{3}$$

where σ_o is the surface tension of the oil.

The surface tension of distilled water is expressed as follows in Equation (4):

$$\sigma_w = 74.33 - 0.127t \tag{4}$$

where σ_w is the surface tension of water.

The power system pumped gas, oil, and water into the pipe. The system consisted of oil and water pumps, an oil tank, a water tank, an air compressor, and a double-Y junction fitting mixer. The metering system was composed of flowmeters and a water holdup instrument. The mixing line comprised a stainless-steel pipe section and a plexiglass pipe section with an inner diameter of 50 mm. The flow pattern development section was a horizontal stainless-steel pipe with a length of 10 m, which was convenient for the full development of the multiphase flow pattern in the pipe. The flow pattern observation section was installed at the end of the flow pattern development section, which was a plexiglass pipe, so that the flow pattern in the pipe could be easily observed. The flow observation sections were a 1-m-long horizontal transparent pipe and a 2-m-long vertical transparent pipe. Through this arrangement, the horizontal and vertical flow experiments could be carried out simultaneously.

The oil, water, and gas were pumped into the pipe from separate storage tanks. The gas was supplied by an air compressor (GA37VSDAP-13, with a capacity of 120.8 L/s) to the gas buffer tank to stabilize its pressure. The volume flow rate was regulated using an orifice flowmeter (EJA115, measurement range of 0.078–94.2 Nm^3/h) or a vortex flowmeter (DY015-DN15, measurement range of 30–275 Nm^3/h), depending on the flow range. A centrifugal pump (QABP160M2A, ABB) with a capacity of 12.5 m^3/h was used for the water phase, and a 6.99-KW gear pump (SNH440) with a capacity of 17 m^3/h and an accuracy of $\pm 0.1\%$ was used for the oil phase. The volumetric flow rates of the oil and water phases were measured by a mass flowmeter (CMF100, Micro Motion), with an accuracy of $\pm 0.1\%$. The gas, oil, and water phases entered the double-Y junction fitting mixer from the upper, middle, and lower layers of the mixer, respectively. A schematic diagram of the double-Y junction fitting mixer is shown in Figure 2. The well-mixed three-phase flow passed through the test section and then flowed back to the separating tank, in which the gas escaped to the atmosphere and the oil and water flowed into the oil and water tanks, respectively.

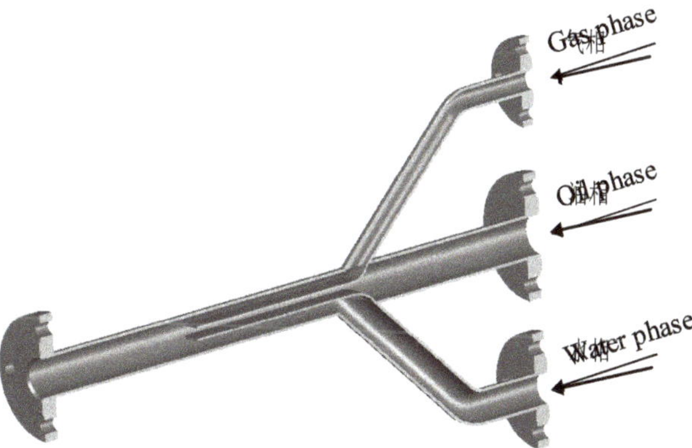

Figure 2. Schematic of the double-Y junction fitting mixer.

When the pressure drop in the pipe remained constant, it was deemed that a steady state of the system had been reached. Data for a period of 300 s were recorded. The pressure drop was measured by a differential pressure transducer placed in the return section. A digital video was used for flow pattern identification. The cross-sectional average water holdup α_w (in situ volume fraction of water) was recorded by two water holdup instruments placed in the horizontal and vertical sections of the test pipe. The water holdup instrument was equipped with a conductance probe having a sampling frequency of 1 Hz. A photograph of the instrument is shown in Figure 3. Two pairs of conductance probes were regularly distributed in the middle of the stainless steel pipe. Each probe was comprised of two parallel brass rods. When alternating current flowed between two probes, the conductance probe measured the voltage between the two ends of the conductor, which reflects the mean conductivity of the mixture in the pipe [20]. Because the conductivities of oil and gas are weak, the voltage values measured when the pipe was filled with pure oil or gas were basically the same. Calibration of the water holdup instrument was performed by measuring the transmitted conductivity for single-phase gas, oil, and water. The calibration information was used to calculate the water holdup for three-phase flows. The measured voltage values are denoted by V_o, V_w, and V_g when the pipe is fully filled with the pure oil, water, and gas, respectively.

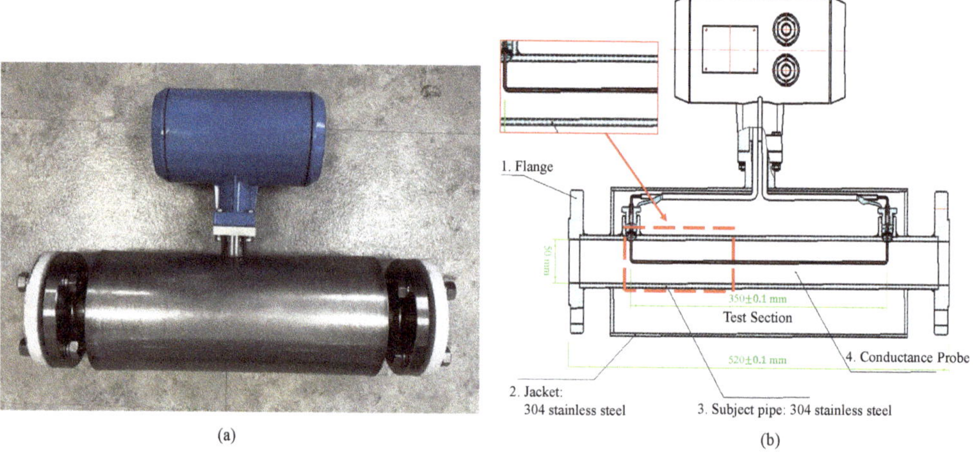

Figure 3. (**a**) Photograph of the water holdup instrument designed for the experiments. (**b**) The internal structure of the water holdup instrument.

We inferred that the mean voltage V_{exp} measured across the flow stream would be a function of the form (Equation (5)):

$$V_{exp} = f(\alpha_w, \alpha_o, V_w, V_o, V_g) \tag{5}$$

The volume fraction of the three phases satisfies the following relation (Equation (6)):

$$\alpha_o + \alpha_w + \alpha_g = 1 \tag{6}$$

Equation (5) can be made dimensionless by using V_w as the characteristic voltage, yielding the following (Equation (7)):

$$\frac{V_{exp}}{V_w} = f(\alpha_w, \alpha_o, \frac{V_0}{V_w}, \frac{V_g}{V_w}) \tag{7}$$

As a classic method, the conductivity measurement method has been widely studied and applied due to its simple structure, convenient installation, and fast response. However, the equivalent conductivity of multiphase fluid is not only related to phase holdup but also affected by flow pattern or phase distribution [21]. This issue has not been solved for more complex flow patterns since the first work by Bruggeman [22]. In general, the mean voltage V_{exp} measured across the flow stream can be assumed to be a linear superposition of the corresponding parameters of each phase for simplicity. Achwal and Stepanek [23,24] used a conductance probe to measure the liquid holdup in a gas–liquid system. As they pointed out, the electroconductivity of a liquid system is proportional to the cross-sectional area of the conducting liquid. Thus, the conductivity should be proportional to the liquid holdup. The same procedure was also adopted by Begovich and Watson [25]. Du et al. [7]

measured the water holdup using a vertical multiple electrode array conductance sensor in vertical upward oil–water flow. According to the experiments, the mean voltage for mixed fluid and oil holdup showed a good linear relationship. Based on the above considerations, the mean voltage V_{exp} can be expressed as follows in Equation (8):

$$\frac{V_{exp}}{V_w} = \alpha_0 \frac{V_0}{V_w} + (1 - \alpha_w - \alpha_o)\frac{V_g}{V_w} + \alpha_w \quad (8)$$

Since the voltage V_o is equal to V_g, the above formula can be further simplified as follows in Equation (9):

$$\frac{V_{exp}}{V_w} = (1 - \alpha_w)\frac{V_o}{V_w} + \alpha_w \quad (9)$$

Thus, the relationship between the water holdup α_w and the mean voltage V_{exp} can be expressed as follows:

$$\alpha_w = \frac{V_0 - V_{exp}}{V_o - V_w} \quad (10)$$

The measured voltage values when the pipe is fully filled with the pure water and pure oil in each group of experiments were different. Therefore, V_o and V_w were measured for each set of experimental conditions when the voltage signal was converted into a water holdup using Equation (10). The dimensionless electrical signal of the water holdup instrument ensured that the measured data of the horizontal and vertical devices in the same group of experiments could be compared, which was also convenient for the comparison of experimental data of different groups.

3. Analysis of Oil–Water–Gas Three-Phase Flow

An oil–water–gas three-phase flow is highly complex and related to pipe geometry, fluid properties, and fluid flow rates. An oil–water–gas three-phase flow can be regarded as a special kind of gas–liquid two-phase flow, especially at high flow rates, where the oil and water are well mixed and form a homogeneous dispersion. The clear identification of the oil and the water phases is difficult in these cases [8]. The methods and theories developed for gas–liquid two-phase flows can be used as the basis for the investigation of oil–water–gas three-phase flows [14]. The liquid mixture properties, such as the viscosity and density, depend on the ratio of the superficial velocity of oil to that of water. The mixture density is defined as follows in Equation (11):

$$\rho_m = \varepsilon_w \rho_w + \varepsilon_o \rho_o \quad (11)$$

where ρ_w is the water density. ε_w and ε_o are the input water and oil cuts, respectively, defined as the ratio of each phase flow rate to the mixture flow rate (Equations (12) and (13)):

$$\varepsilon_w = \frac{Q_w}{Q_w + Q_o} = \frac{u_{sw}}{u_{sm}} \quad (12)$$

$$\varepsilon_o = \frac{Q_o}{Q_w + Q_o} = \frac{u_{so}}{u_{sm}} \quad (13)$$

where Q is the volume flow rate of each phase, and u_s is the superficial velocity, which is defined as $u_s = Q/A$, A is the cross-sectional area of the pipe, $A = \pi D^2/4$, and D is the inner pipe diameter.

An oil–water mixture is a non-Newtonian fluid, and its viscosity is called the apparent viscosity. In general, the viscosity of a liquid mixture varies significantly with the spatial distribution of the two phases, the viscosity of each phase, the temperature, and the pressure. It is difficult to include all these factors in any theoretical expression of the apparent viscosity. Some scholars have suggested using a calculation method similar to that for the densities of oil–water mixtures to calculate the apparent viscosity [11,26,27]. Over the range of superficial velocities considered here, the oil and water were well mixed, and the liquid phases appeared "milky." In this study, it was assumed that the liquid mixture viscosity depended on the water and oil fractions for convenience. The mixture viscosity is given as follows in Equation (14):

$$\mu_m = \varepsilon_w \mu_w + \varepsilon_o \mu_o \tag{14}$$

where μ_m is the water viscosity.

Dimensional analysis is a useful tool to obtain the coupling effect of the controlling factors on the two-phase flow behavior. The factors affecting the gas–liquid two-phase flow are listed as follows:

- Gas phase: density ρ_g, viscosity μ_g.
- Liquid mixture: density ρ_m, viscosity μ_m, interfacial tension σ.
- Geometric parameter: inner pipe diameter D.
- Boundary condition: superficial velocity of gas u_{sg}, superficial velocity of the liquid mixture u_{sm}.
- Gravitational acceleration: g.

The final steady-state of the gas–liquid flow system, characterized by the water holdup α_w and the flow pattern, is a function of the above control parameters (Equation (15)):

$$\begin{cases} \alpha_w = f(\rho_g, \mu_g, u_{sg}, \sigma, \rho_m, \mu_m, u_{sm}; D, g), \ \varepsilon_w \neq 0 \\ flow\ pattern = f(\rho_g, \mu_g, u_{sg}, \sigma, \rho_m, \mu_m, u_{sm}; D, g) \end{cases} \tag{15}$$

The above formula can be nondimensionalized as follows in Equation (16):

$$\begin{cases} \alpha_w = f\left(\frac{u_{sg}}{u_{sm}}, \frac{\rho_m u_{sm} D}{\mu_m}, \frac{u_{sm}^2}{gD}, \frac{|\rho_m - \rho_g| g D^2}{\sigma}, \frac{\mu_g}{\mu_m}, \frac{\rho_g}{\rho_m}\right), \ \varepsilon_w \neq 0 \\ flow\ pattern = f\left(\frac{u_{sg}}{u_{sm}}, \frac{\rho_m u_{sm} D}{\mu_m}, \frac{u_{sm}^2}{gD}, \frac{|\rho_m - \rho_g| g D^2}{\sigma}, \frac{\mu_g}{\mu_m}, \frac{\rho_g}{\rho_m}\right) \end{cases} \tag{16}$$

where u_{sg}/u_{sm} is the gas-to-liquid superficial velocity ratio, $\rho_m u_{sm} D/\mu_m$ is the Reynolds number Re_m, u_{sm}^2/gD is the Froude number Fr_m, $(|\rho_m - \rho_g| g D^2)/\sigma$ is the Eötvös number Eo, which represents the ratio of the buoyancy force to the surface tension force, μ_g/μ_m is the viscosity ratio, and ρ_g/ρ_m is the density ratio. In this study, the input water cut ε_w ranged from 0% to 100%, corresponding to μ_m values from 1 to 32 mPa/s and ρ_m values from 843 to 998.2 kg/m3. Because the density and viscosity of the gas were much lower than that of the liquid mixture, the effect of variations of Eo, ρ_g/ρ_m, and μ_g/μ_m were not considered in this study. Thus, Equation (16) can be simplified as follows:

$$\begin{cases} \alpha_w = f\left(\frac{u_{sg}}{u_{sm}}, Re_m, Fr_m\right), \ \varepsilon_w \neq 0 \\ flow\ pattern = f\left(\frac{u_{sg}}{u_{sm}}, Re_m, Fr_m\right) \end{cases} \tag{17}$$

The input water cut ε_w is an important parameter for oil pipeline transportation. The effect of ε_w on the final steady-state is reflected in the Reynolds number Re_m Tests were conducted for different values of the dimensionless parameters in Equation (17) to associate the observed flow patterns with the measured water holdup values for the horizontal and vertical sections of the pipe. All the experimental values of u_{sg}, u_{sg}, and ε_w and the corresponding dimensionless numbers are given in Table 2.

Table 2. Parameter values of flow conditions imposed during the oil–water–gas pipe flow experiments.

Test	ε_w	u_{sg}/u_{sm}	$Re_m = \rho_m u_{sm} D/u_{sm}$	$Fr_m = u_{sm}^2/gD$
1	1.0	0.25, 0.50, 1, 2, 3, 4, 5, 6	11,303	0.105
2	0.9	0.25, 0.50, 1, 2, 3, 4, 5, 6	2714	0.105
3	0.8	0.25, 0.50, 1, 2, 3, 4, 5, 6	1521	0.105
4	0.7	0.25, 0.50, 1, 2, 3, 4, 5, 6	1046	0.105
5	0.6	0.25, 0.50, 1, 2, 3, 4, 5, 6	791	0.105
6	0.5	0.25, 0.50, 1, 2, 3, 4, 5, 6	632	0.105
7	0.4	0.25, 0.50, 1, 2, 3, 4, 5, 6	523	0.105
8	0.3	0.25, 0.50, 1, 2, 3, 4, 5, 6	444	0.105
9	0.2	0.25, 0.50, 1, 2, 3, 4, 5, 6	384	0.105
10	0.1	0.25, 0.50, 1, 2, 3, 4, 5, 6	336	0.105
11	0.0	0.25, 0.50, 1, 2, 3, 4, 5, 6	298	0.105
12	1.0	0.25, 0.50, 1, 2, 3, 4, 5, 6	16,248	0.216
13	0.9	0.25, 0.50, 1, 2, 3, 4, 5, 6	3901	0.216
14	0.8	1, 2, 3, 4, 5, 6	2187	0.216
15	0.7	0.25, 0.50, 1, 2, 3, 4, 5, 6	1504	0.216
16	0.6	0.25, 0.50, 1, 2, 3, 4, 5, 6	1137	0.216
17	0.5	0.25, 0.50, 1, 2, 3, 4, 5, 6	908	0.216
18	0.4	0.25, 0.50, 1, 2, 3, 4, 5, 6	752	0.216
19	0.3	0.25, 0.50, 1, 2, 3, 4, 5, 6	638	0.216
20	0.2	0.25, 0.50, 1, 2, 3, 4, 5, 6	551	0.216
21	0.1	0.25, 0.50, 1, 2, 3, 4, 5, 6	484	0.216
22	0.0	0.25, 0.50, 1, 2, 3, 4, 5, 6	429	0.216
23	1.0	0.25, 0.50, 1, 2, 3, 4	21,193	0.368
24	0.9	0.25, 0.50, 1, 2, 3, 4	5089	0.368
25	0.8	0.25, 0.50, 1, 2, 3, 4	2852	0.368
26	0.7	0.25, 0.50, 1, 2, 3, 4	1962	0.368
27	0.6	0.25, 0.50, 1, 2, 3, 4	1483	0.368
28	0.5	0.25, 0.50, 1, 2, 3, 4	1185	0.368
29	0.4	0.25, 0.50, 1, 2, 3, 4	980	0.368
30	0.3	0.25, 0.50, 1, 2, 3, 4	832	0.368
31	0.2	0.25, 0.50, 1, 2, 3, 4	719	0.368
32	0.1	0.25, 0.50, 1, 2, 3, 4	631	0.368
33	0.0	0.25, 0.50, 1, 2, 3, 4	559	0.368

4. Results and Discussion

4.1. Flow Pattern Maps

The identification of the flow patterns was based on both visual observations from the video camera and the PDF of the instantaneous cross-sectional water holdup measured by the water holdup instrument. According to the classification by Weisman [28], the flow patterns for a gas–liquid flow were classified into bubbly, slug, plug, annular, stratified, and disperse flows in the horizontal pipe and bubbly, slug, churn, annular, and disperse flows in the vertical pipe. Figure 4 shows the schematic representations of the horizontal and vertical gas–liquid flow patterns defined by Weisman [28]. Over the range of superficial velocities considered here, plug, slug, and annular flows were observed in the horizontal section. In the vertical section, slug and churn flows were observed.

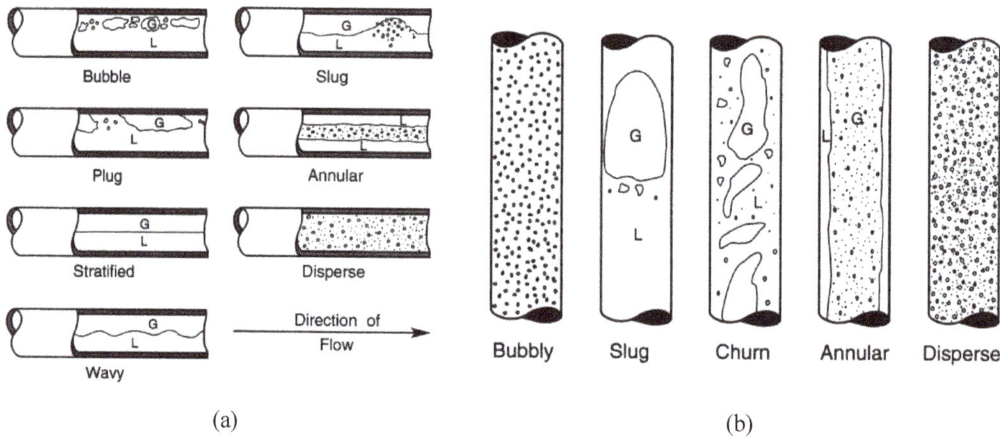

Figure 4. Schematic representation of the (**a**) horizontal and (**b**) vertical gas–liquid flow patterns defined by Weisman [28].

Figure 5 shows examples of the four main flow patterns that were observed in this study in both the horizontal and vertical sections in a continuous transportation pipe. Figure 5a presents the plug flow pattern, where small bubbles gathered in the upper part of the pipe and formed large bubbles, and there were almost no small bubbles between the large bubbles. Figure 5b shows the annular flow pattern, where a liquid film with a certain thickness formed between the gas column and the pipe wall. Figure 5c shows the slug flow pattern, where gas pockets were separated by slugs of liquid with dispersed small bubbles. Figure 5d shows the churn flow pattern; the flow was similar to the slug flow but without clear phase separation or structure. Figure 5 shows that the oil and water phases were well mixed with each other. Based on the experimental observations, the assumption that the oil–water phase was simplified as a liquid mixture was reasonable.

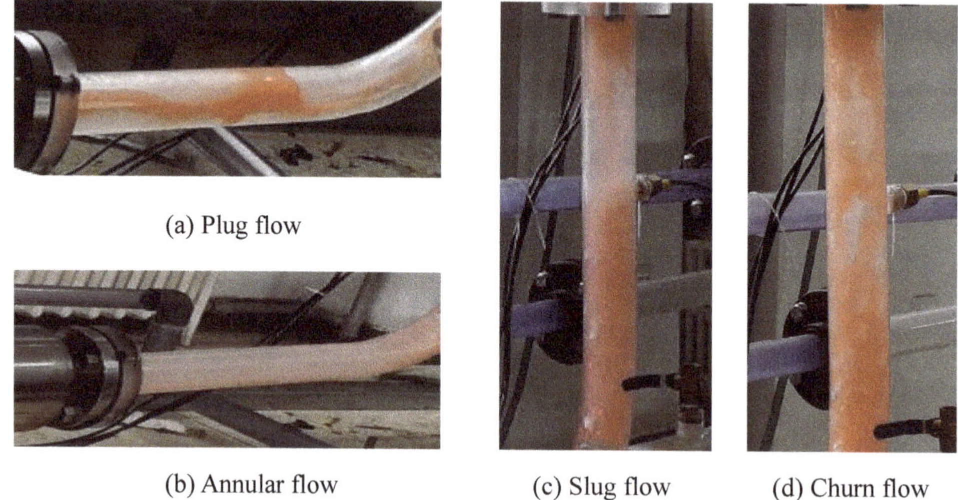

Figure 5. Instantaneous flow patterns observed in the experiments in the horizontal and vertical sections: (**a**) plug flow, Fr_m = 0.368, u_{sg}/u_{sm} = 0.25, ε_w = 0.5; (**b**) annular flow Fr_m = 0.105, u_{sg}/u_{sm} = 0.50, ε_w = 0.5; (**c**) slug flow, Fr_m = 0.216, u_{sg}/u_{sm} = 6, ε_w = 0.1; (**d**) churn flow, Fr_m = 0.368, u_{sg}/u_{sm} = 4, ε_w = 0.1.

Figures 6 and 7 show the typical time-series data of the cross-sectional average water holdup α_w and the corresponding PDF as a function of the water holdup for different superficial velocities in both the horizontal and vertical sections [4]. Figure 6 shows the data for Fr_m = 0.368, u_{sg}/u_{sm} = 0.50, and ε_w = 0.8. In the horizontal section, plug flow was observed. With the increase in the gas flow rate, small bubbles coalesced in the upper part of the pipe to form large bubbles. Few small bubbles existed between the large bubbles. The plug flows were characterized by two peaks in the PDF at α_w = 0.55 and 0.75 in this case. The two peaks corresponded to the bubble region and the plug region, respectively. In the vertical section, slug flow was observed under the same experimental conditions. Taylor bubbles and liquid slugs between two adjacent Taylor bubbles formed slug flows. Many small bubbles were distributed between the Taylor bubbles, which is referred to as the wake region [29]. In this study, slug flows were characterized by three peaks in the PDF at α_w = 0.45, 0.60, and 0.80. The regions at α_w = 0.45, 0.60, and 0.80 corresponded to the Taylor bubble, wake, and liquid slug regions, respectively.

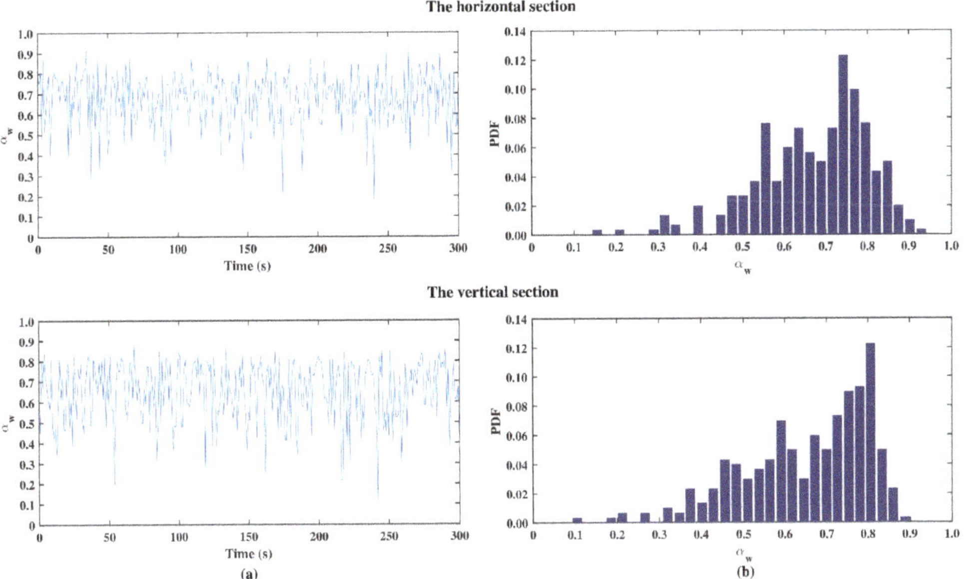

Figure 6. Typical evolution of the cross-sectional average water holdup α_w and the probability density function (PDF) of the fluctuations in the volume fraction for oil–water–gas pipe flow (Fr_m = 0.368, u_{sg}/u_{sm} = 0.50, and ε_w = 0.8). Plug flow in the (**a**) horizontal section and slug flow in the (**b**) vertical section.

Figure 7 shows the result for Fr_m = 0.105, u_{sg}/u_{sm} = 6, and ε_w = 0.5. Annular flow was observed in the horizontal section in this case. The PDF showed a single narrow peak at a low input water holdup (α_w = 0.02). The point at which the maximum occurred in the annular flow PDF represents the volume fraction of the liquid film surrounding the gas column. In the vertical section, churn flow was observed, and the associated PDF diagram showed a single peak that was similar to the PDF associated with the annular flow at a low input water cut. However, the distribution range of the single peak was greater than that in the annular flow, suggesting that significantly severely agitated mixing occurred.

Flow pattern maps were developed for the oil–water–gas pipe flow in the horizontal and vertical sections, which are shown in Figures 8 and 9. In these figures, the x-, y-, and z-axes corresponded to the dimensionless numbers Fr_m, u_{sg}/u_{sm}, and Re_m, respectively. Fr_m values of 0.105, 0.216, and 0.368 were investigated, and Re_m and u_{sg}/u_{sm} were in the ranges of 298–21,193 and 0.25–6, respectively. The flow pattern maps depended on the pipe size and the size of the gas inclusions. The flow pattern maps in this study were constructed for $D = 50$ mm.

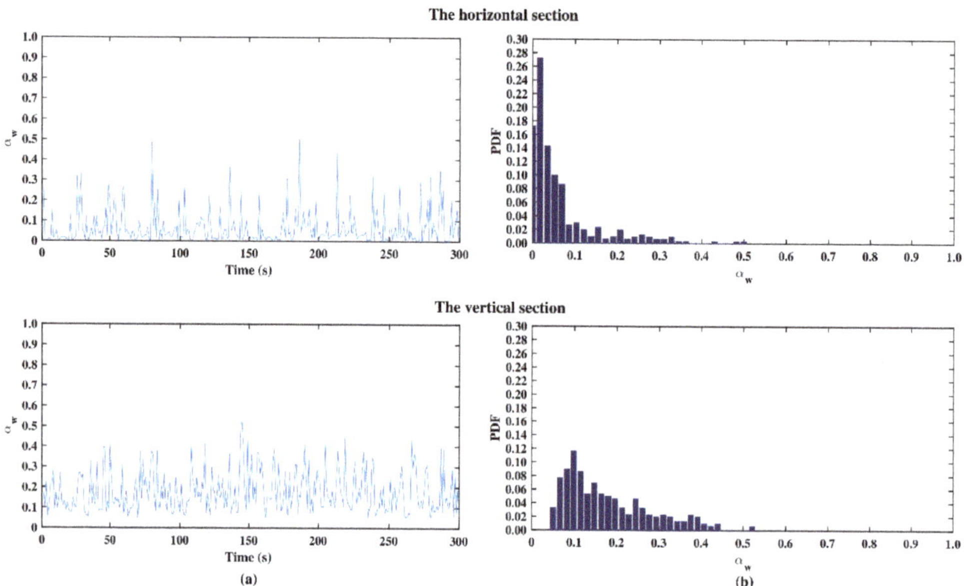

Figure 7. Typical evolution of the cross-sectional average water holdup α_w and the probability density function (PDF) of the fluctuations in the volume fraction for oil–water–gas pipe flow ($Fr_m = 0.105$, $u_{sg}/u_{sm} = 6$, and $\varepsilon_w = 0.5$). Annular flow in the (**a**) horizontal section and churn flow in the (**b**) vertical section.

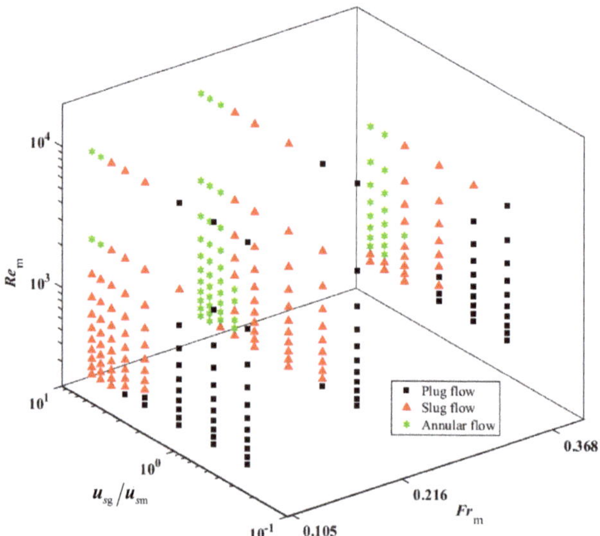

Figure 8. Flow pattern maps in the horizontal section, in which the dimensionless number Re_m ranged from 298 to 21,193, the dimensionless number u_{sg}/u_{sm} ranged from 0.25 to 6, and the dimensionless number Fr_m was fixed at 0.105, 0.216, and 0.368.

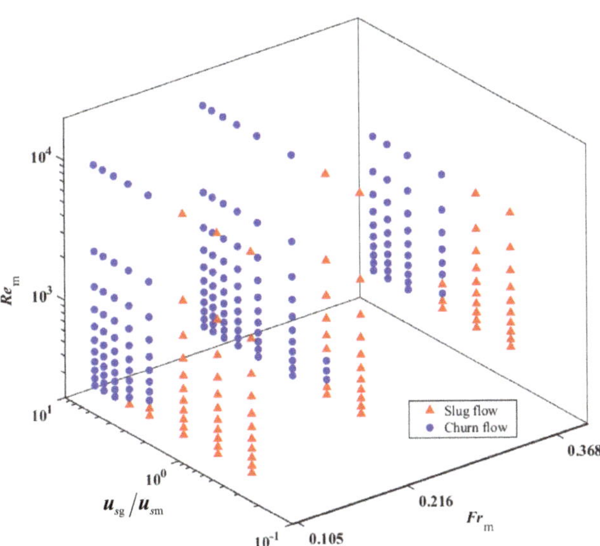

Figure 9. Flow pattern maps in the vertical section, in which the dimensionless number Re_m ranged from 298 to 21,193, the dimensionless number u_{sg}/u_{sm} ranged from 0.25 to 6, and the dimensionless number Fr_m was fixed at 0.105, 0.216, and 0.368.

In the horizontal section (Figure 8), at low u_{sg}/u_{sm} values, plug flow was observed. For a given Re_m and increasing u_{sg}/u_{sm}, the gas content increased, converting plug flow to slug or annular flow. At high Re_m and u_{sg}/u_{sm}, annular flow was identified. By increasing Fr_m from 0.105 to 0.216 and 0.368, the transition boundary between the slug and annular flows moved to the right side of the map. In other words, by increasing Fr_m, the annular flow regime zone was expanded. The experimental flow pattern map in the vertical section is shown in Figure 9. Only slug and churn flows were observed in the experimental conditions. A further increase in u_{sg}/u_{sm} converted the slug flow to churn flow. Compared with the horizontal section, plug flow was replaced by slug flow at low u_{sg}/u_{sm} values, and slug flow was replaced by churn flow at high u_{sg}/u_{sm} values. By increasing Fr_m, the transition boundary between the slug and churn flows moved to the right side of the map, i.e., lower u_{sg}/u_{sm} values.

The comparison of Figures 8 and 9 shows that the plug flow in the horizontal section corresponded to the slug flow in the vertical section, and the slug flow in the horizontal section corresponded to the churn flow in the vertical section. At low u_{sg}/u_{sm} values, the vertical section exhibited slug flow, while in the horizontal section, bubbles could float to the upper part of the pipe to form plug flow due to buoyancy. At high u_{sg}/u_{sm} values, the horizontal section exhibited slug flow due to the inertial forces of the bubbles being very large, while in the vertical section, the bubbles broke up and formed churn flow instead.

4.2. Water Holdup

In this study, the water holdup instrument was employed to measure the steady-state water holdup in the horizontal and vertical sections of the pipe. However, the measured holdup data for the horizontal section at high u_{sg}/u_{sm} values were considered to be less reliable because the conductance probe was placed parallel to the flow direction. Therefore, only the water holdup measured in the vertical section is considered in the following discussion. In the following work, the structure of the water holdup instrument can be improved to get more accurate experimental data in the horizontal section.

Figure 10 illustrates the relationship between the measured water holdup α_w and u_{sg}/u_{sm} for different Re_m values. Fr_m values of 0.105, 0.216, and 0.368 were investigated.

The increase in u_{sg}/u_{sm} led to a decrease in the water holdup. For a given u_{sg}/u_{sm}, the water holdup decreased with the decrease in Re_m. Moreover, for large u_{sg}/u_{sm}, the amount of decrease of the water holdup became smaller and smaller until it remained unchanged. For low Re_m (corresponding to input water cuts of 0.1 and 0.2), the error of the measured water holdup was large, causing the water holdup to show different trends. According to Figure 10a–c, it is noted that the trend between α_w and u_{sg}/u_{sm} remained essentially the same, regardless of the value of Fr_m. The value of u_{sg}/u_{sm} had an important influence on the water holdup α_w.

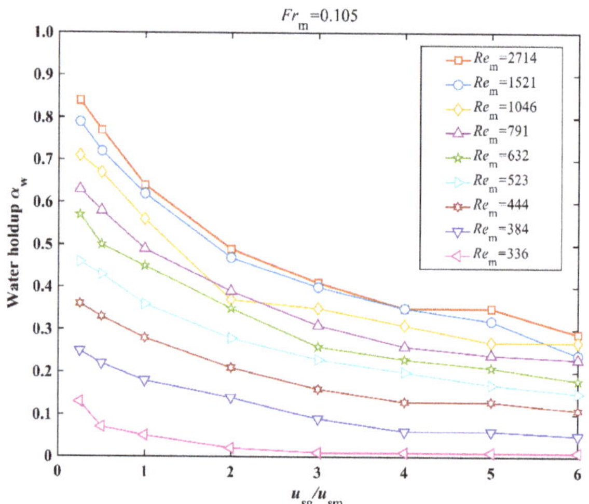

(**a**) Water holdup α_w versus u_{sg}/u_{sm} for $Fr_m = 0.105$

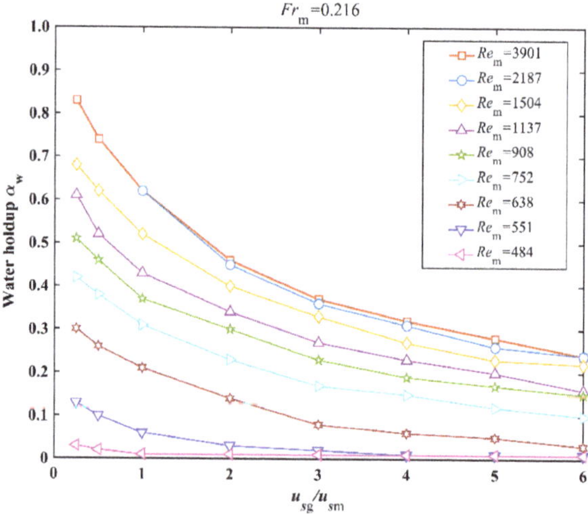

(**b**) Water holdup α_w versus u_{sg}/u_{sm} for $Fr_m = 0.216$

Figure 10. *Cont.*

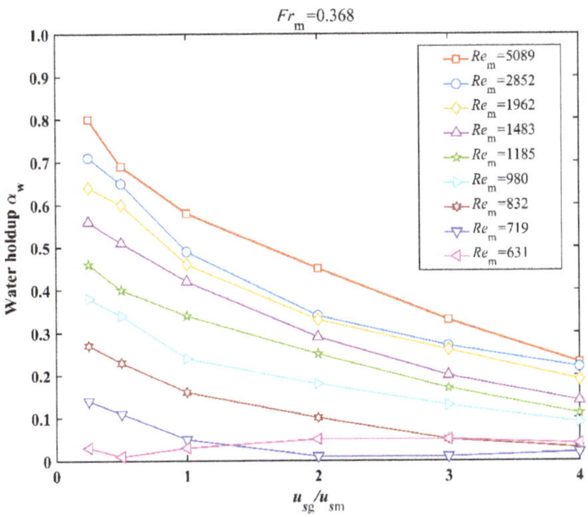

(c) Water holdup α_w verses u_{sg}/u_{sm} for $Fr_m = 0.368$

Figure 10. Effect of u_{sg}/u_{sm} on water holdup α_w in the vertical section for different Re_m values. The input water cut ε_w ranged from 10 to 90%. Fr_m was fixed at (**a**) 0.105, (**b**) 0.216, and (**c**) 0.368.

Figure 11 shows the water holdup α_w as a function of Re_m for different u_{sg}/u_{sm}. The data for three different Fr_m values (0.105, 0.216, and 0.368) are presented. The increase in Re_m led to an increase in the water holdup. For a given Re_m, the water holdup decreased with the increase in u_{sg}/u_{sm}. The results were reasonably close to the results shown in Figure 10 and exhibited similar trends. By analogy to Figure 11a–c, it is noted that the trend between α_w and Re_m remained essentially the same, regardless of the values of Fr_m. The Reynolds number Re_m is also a key parameter affecting the flow behavior of the oil–water–gas three-phase flow.

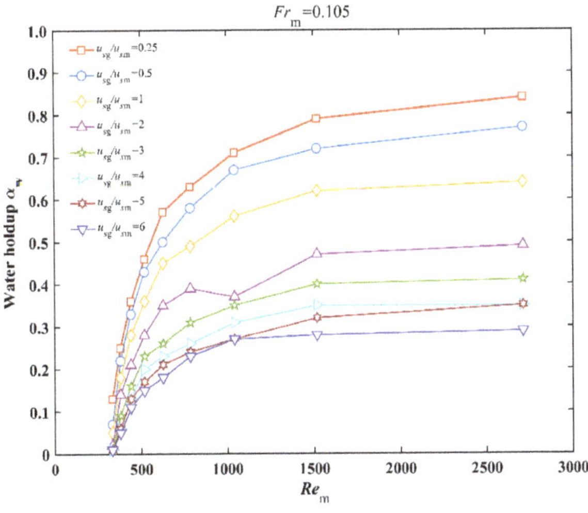

(a) Water holdup α_w verses Re_m for $Fr_m = 0.105$

Figure 11. *Cont.*

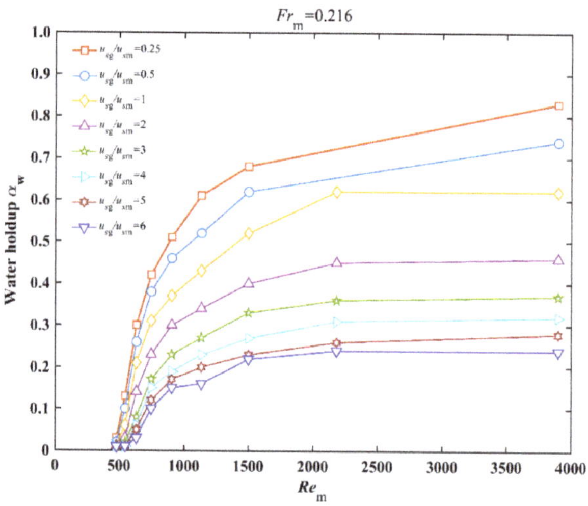

(**b**) Water holdup α_w verses Re_m for $Fr_m = 0.216$

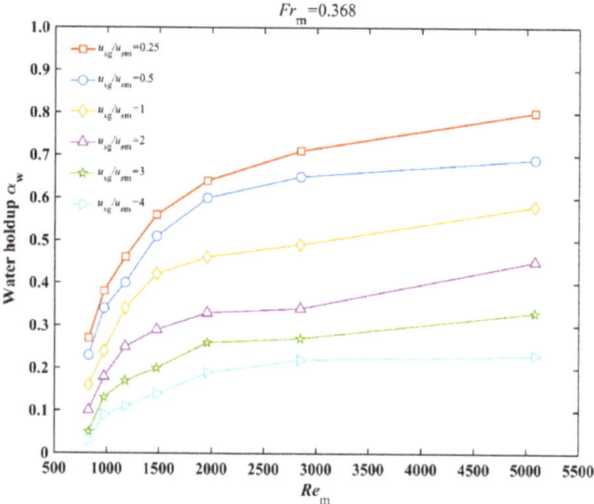

(**c**) Water holdup α_w verses Re_m for $Fr_m = 0.368$

Figure 11. Effect of Re_m on water holdup α_w in the vertical section for different u_{sg}/u_{sm} values. The input water cut ε_w ranged from 10 to 90%. Fr_m was fixed at (**a**) 0.105, (**b**) 0.216, and (**c**) 0.368.

According to Equation (17), the water holdup in the vertical section α_w is a function of Re_m, Fr_m, and u_{sg}/u_{sm}. The relationship between α_w and the corresponding dimensionless numbers can be written as the following power-law correlation in Equation (18):

$$\alpha_w = a(Re_m)^b(Fr_m)^c\left(\frac{u_{sg}}{u_{sm}}\right)^d, \ \varepsilon_w \neq 0 \tag{18}$$

where a is a pre-factor, and b, c, and d are the fitted exponents. The experimental data were fitted to obtain the best correlation through the least-squares method. The fitting result is as follows:

$$\alpha_w = 0.06(Re_m)^{0.20}(Fr_m)^{-0.26}\left(\frac{u_{sg}}{u_{sm}}\right)^{-0.34}, \ \varepsilon_w \neq 0 \tag{19}$$

i.e., $a = 0.06$, $b = 0.20$, $c = -0.26$, and $d = -0.34$. The data from which the above correlation was derived were collected for the following parameter ranges: $0.25 \leq u_{sg}/u_{sm} \leq 6$, $336 \leq Re_m \leq 5089$, and $0.105 \leq Fr_m \leq 0.368$. The coefficient of determination R^2 of Equation (19) was 0.83. The standard deviation (SD) of the predicted value was 8.5%, which was calculated by the following correlation in Equation (20) [30]:

$$SD = \sqrt{\frac{1}{n-1}\sum_{k=1}^{n}\left(\frac{(\alpha_w)_{pred} - (\alpha_w)_{exp}}{(\alpha_w)_{exp}}\right)^2} \quad (20)$$

Figure 12 depicts a comparison of the predicted water holdup and the experimental data. The square, circular, and triangle symbols represent the data for $Fr_m = 0.105$, 0.216, and 0.368, respectively. The developed correlation (Equation (19)) demonstrates a good agreement with the experimental data. It is noted that the result measured by the water holdup instrument was smaller than the real water holdup α_w in the pipe when the input water cut ε_w was less than 0.3. This was the reason for the predicted water holdup being away from the correlation reference line in the bottom-left part of the diagram. This discrepancy can be solved by improving the accuracy of the water holdup instrument in future research. Moreover, there was no evident structure to the degree of correlation with respect to the values of the Froude number Fr_m of the oil–water mixture. This proved the applicability of Equation (19) for different Froude numbers. The power-law water holdup correlation is dimensionless, and it can be extended to other conditions. As can be seen in Figure 13, the classical Beggs-Brill empirical model [31] and the developed correlation (Equation (19)) in this study are used to predict the experimental water holdup data. The performance of Equation (19) is better than that of the Beggs-Brill model.

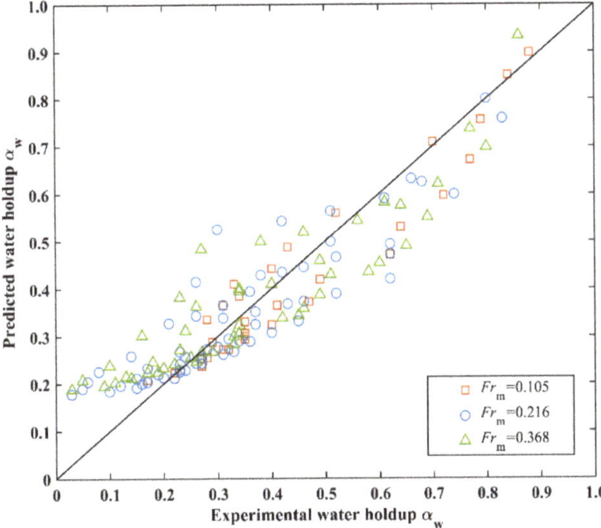

Figure 12. Comparison between the predicted water holdup and the experimental data for the oil–water–gas three-phase flow. The input water cut ε_w ranged from 10 to 90%. The brown squares corresponded to $Fr_m = 0.105$, the blue circles corresponded to $Fr_m = 0.216$, and the green triangles corresponded to $Fr_m = 0.368$.

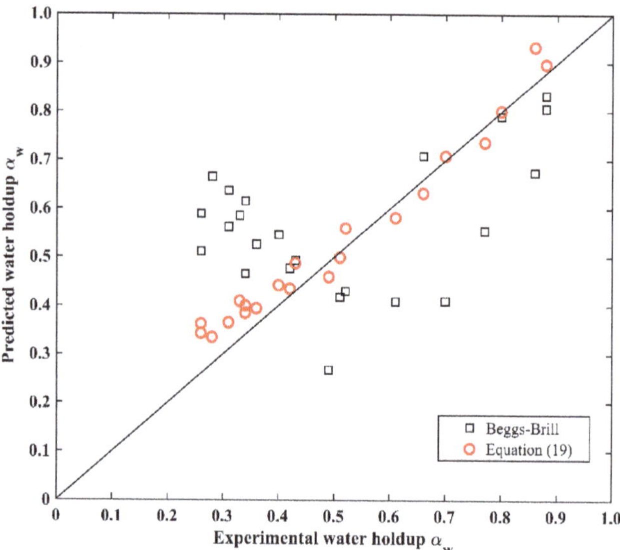

Figure 13. Comparison of different models predicting water holdup with the experimental data for the gas–water two-phase flow. The input water cut ε_w is equal to 100%. The red circles corresponded to Equation (19), the black squares corresponded to the Beggs-Brill empirical model [31].

5. Conclusions

Due to the presence of water in oil wells or the injection of water into a well to increase the oil production, the pipe flow is in the form of oil–gas two-phase flows and oil–gas–water three-phase flows during oil exploitation and transportation. The change in the flow pattern and water holdup during oil pipeline transportation is important for the proper design and operation of pipelines. The change of the flow pattern has a significant effect on the pressure drop in the pipeline, and the calculation of the water holdup helps to predict the quantity of oil in petroleum pipelines. For example, the scaling and corrosion of a pipe can be prevented by controlling the proper flow pattern. In the long-distance multiphase transportation of high-viscosity crude oil, the flow can be artificially controlled as dispersed or annular flow with water as the continuous phase to reduce pressure drop losses. The judgment of the flow pattern and water holdup can also provide a quantitative basis for the separation of oil and water. Thus, a better understanding of the flow patterns and water holdup of the oil–water–gas three-phase flow is beneficial for the proper design and operation of pipelines. Meanwhile, it is also an essential topic in multiphase hydraulics and water resource management.

Oil–water–gas three-phase flow experiments were conducted in a pipe consisting of a horizontal section and a vertical section simultaneously. The newly designed water holdup instrument equipped with a conductance probe was used to measure the cross-sectional average water holdup. The flow behavior of the oil–water–gas three-phase flow was studied using dimensional analysis. The oil–water–gas three-phase flow was simplified as the two-phase flow of a gas and a liquid mixture. The effects of the dimensionless numbers u_{sg}/u_{sm}, Re_m, and Fr_m on the multiphase flow in the pipe were analyzed and discussed.

New flow pattern maps of the three-phase flow in terms of Re_m and u_{sg}/u_{sm} were proposed over the range of superficial velocities considered. Plug, slug, and annular flows were observed in the horizontal section. Only slug and churn flows were observed in the vertical section. The plug flow was characterized by two peaks in the PDF, and the two peaks corresponded to the bubble region and the plug region, respectively. Slug flow was characterized by three peaks in the PDF, and the three peaks corresponded to the Taylor bubble, wake, and liquid slug regions, respectively. The annular and churn flows were characterized by a single peak in the PDF, and the distribution range of the single peak in the churn flow was greater than that in the annular flow. Moreover, the flow pattern maps described using dimensionless numbers may also be applied to other pipe sizes. Based on the experimental data, a dimensionless power-law water holdup correlation for the oil–water–gas three-phase flow in the vertical section was developed. The predicted water holdup agreed reasonably well with the experimental results.

Author Contributions: Conceptualization: G.R., D.G. and P.L.; Formal analysis: G.R., D.G., P.L., X.Z. and X.L.; Investigation: G.R., P.L., X.Z. and X.L.; Methodology: G.R., X.C., K.S., R.F., L.M. and F.S.; Project administration: G.R., X.C., K.S., R.F., L.M. and F.S.; Supervision: G.R. and D.G.; Validation: P.L.; Visualization: P.L.; Writing—original draft: P.L.; Writing—review and editing: P.L., X.Z. and X.L. All authors have read and agreed to the published version of the manuscript.

Funding: This research received no external funding.

Institutional Review Board Statement: Not applicable.

Informed Consent Statement: Not applicable.

Data Availability Statement: Not applicable.

Conflicts of Interest: The authors declare no conflict of interest.

References

1. Brauner, N. The prediction of dispersed flows boundaries in liquid–liquid and gas–liquid systems. *Int. J. Multiph. Flow* **2001**, *27*, 885–910. [CrossRef]
2. Ibarra, R.; Nossen, J.; Tutkun, M. Two-phase gas–liquid flow in concentric and fully eccentric annuli. Part I: Flow patterns, holdup, slip ratio and pressure gradient. *Chem. Eng. Sci.* **2019**, *203*, 489–500. [CrossRef]
3. Xu, J.Y.; Li, D.H.; Guo, J.; Wu, Y.X. Investigations of phase inversion and frictional pressure gradients in upward and downward oil–water flow in vertical pipes. *Int. J. Multiph. Flow* **2010**, *36*, 930–939. [CrossRef]
4. Jones, O.C.; Zuber, N. The interrelation between void fraction fluctuations and flow patterns in two-phase flow. *Int. J. Multiph. Flow* **1975**, *2*, 273–306. [CrossRef]
5. Hasan, A.; Kabir, C.S. A new model for two-phase oil/water flow: Production log interpretation and tubular calculations. *SPE Prod. Eng.* **1990**, *5*, 193–199. [CrossRef]
6. Liu, Z.; Liao, R.; Luo, W.; Riberio, J.X.F. A new model for predicting liquid holdup in two-phase flow under high gas and liquid velocities. *Sci. Iran.* **2019**, *26*, 1529–1539.
7. Du, M.; Jin, N.D.; Gao, Z.K.; Wang, Z.Y.; Zhai, L.S. Flow pattern and water holdup measurements of vertical upward oil–water two-phase flow in small diameter pipes. *Int. J. Multiph. Flow* **2012**, *41*, 91–105. [CrossRef]
8. Oddie, G.; Shi, H.; Durlofsky, L.J.; Aziz, K.; Pfeffer, B.; Holmes, J.A. Experimental study of two and three phase flows in large diameter inclined pipes. *Int. J. Multiph. Flow* **2003**, *29*, 527–558. [CrossRef]
9. Spedding, P.L.; Donnelly, G.F.; Cole, J.S. Three phase oil–water–gas horizontal co-current flow. *Chem. Eng. Res. Des.* **2005**, *83*, 401–411. [CrossRef]
10. Descamps, M.N.; Oliemans, R.V.A.; Ooms, G.; Mudde, R.F. Experimental investigation of three-phase flow in a vertical pipe: Local characteristics of the gas phase for gas-lift conditions. *Int. J. Multiph. Flow* **2007**, *33*, 1205–1221. [CrossRef]
11. Hanafizadeh, P.; Shahani, A.; Ghanavati, A.; Akhavan-Behabadi, M.A. Experimental investigation of air–water–oil three-phase flow patterns in inclined pipes. *Exp. Therm. Fluid Sci.* **2017**, *84*, 286–298. [CrossRef]
12. Tunstall, R.; Skillen, A. Large eddy simulation of a t-junction with upstream elbow: The role of dean vortices in thermal fatigue. *Appl. Therm. Eng.* **2016**, *107*, 672–680. [CrossRef]
13. Xing, L.; Yeung, H.; Shen, J.; Cao, Y. Numerical study on mitigating severe slugging in pipeline/riser system with wavy pipe. *Int. J. Multiph. Flow* **2013**, *53*, 1–10. [CrossRef]
14. Ghorai, S.; Suri, V.; Nigam, K.D.P. Numerical modeling of three-phase stratified flow in pipes. *Chem. Eng. Sci.* **2005**, *60*, 6637–6648. [CrossRef]
15. Friedemann, C.; Mortensen, M.; Nossen, J. Gas–liquid slug flow in a horizontal concentric annulus, a comparison of numerical simulations and experimental data. *Int. J. Heat Fluid Flow* **2019**, *78*, 108437. [CrossRef]
16. Leporini, M.; Terenzi, A.; Marchetti, B.; Corvaro, F.; Polonara, F. On the numerical simulation of sand transport in liquid and multiphase pipelines. *J. Pet. Sci. Eng.* **2019**, *175*, 519–535. [CrossRef]
17. Pietrzak, M.; Witczak, S. Flow patterns and void fractions of phases during gas–liquid two-phase and gas–liquid–liquid three-phase flow in U-bends. *Int. J. Heat Fluid Flow* **2013**, *44*, 700–710. [CrossRef]
18. Liu, W.; Bai, B. Transition from bubble flow to slug flow along the streamwise direction in a gas–liquid swirling flow. *Chem. Eng. Sci.* **2019**, *202*, 392–402. [CrossRef]

19. Wang, H.Q. Investigation on Flow Characteristics of Oil–Water Two-Phase Flow and Gas–Oil–Water Three-Phase Flow in Horizontal Pipelines. Ph.D. Thesis, China University of Petroleum (East China), Qingdao, China, 2008.
20. Flores, J.G. Characterization of oil-water flow patterns in vertical and deviated wells. *Soc. Pet. Eng.* **1997**, *14*, 102–109.
21. Ji, H.F.; Chang, Y.; Huang, Z.Y.; Wang, B.L.; Li, H.Q. Voidage measurement of gas–liquid two-phase flow based on Capacitively Coupled Contactless Conductivity Detection. *Flow Meas. Instrum.* **2014**, *40*, 199–205. [CrossRef]
22. Bruggeman, D.A.G. Berechnung verschiedener physikalischer Konstanten von heterogenen Substanzen. I. Dielektrizitätskonstanten und Leitfähigkeiten der Mischkörper aus isotropen Substanzen. *Ann. Phys.* **1935**, *416*, 636–664. [CrossRef]
23. Achwal, S.K.; Stepanek, J.B. An alternative method of determining hold-up in gas–liquid systems. *Chem. Eng. Sci.* **1975**, *30*, 1443–1444. [CrossRef]
24. Achwal, S.K.; Stepanek, J.B. Holdup profiles in packed beds. *Chem. Eng. J.* **1976**, *12*, 69–75. [CrossRef]
25. Begovich, J.M.; Watson, J.S. An electroconductivity technique for the measurement of axial variation of holdups in three-phase fluidized beds. *AIChE J.* **1978**, *24*, 351–354. [CrossRef]
26. Huang, S.F.; Zhang, B.D.; Lu, J.; Wang, D. Study on flow pattern maps in hilly-terrain air–water–oil three-phase flows. *Exp. Therm. Fluid Sci.* **2013**, *47*, 158–171. [CrossRef]
27. Hapanowicz, J.; Troniewski, L. Two-phase flow of liquid–liquid mixture in the range of the water droplet pattern. *Chem. Eng. Process.* **2002**, *41*, 165–172. [CrossRef]
28. Weisman, J. Two-phase flow patterns. In *Handbook of Fluids in Motion*; Cheremisinoff, N.P., Gupta, R., Eds.; Chapter 15; Ann Arbor Science Publishers: Ann Arbor, Michigan, USA, 1983; pp. 409–425.
29. Zhang, S.; Wang, Z.; Sun, B.; Yuan, K. Pattern transition of a gas–liquid flow with zero liquid superficial velocity in a vertical tube. *Int. J. Multiph. Flow* **2019**, *118*, 270–282. [CrossRef]
30. Al-Wahaibi, T. Pressure gradient correlation for oil–water separated flow in horizontal pipes. *Exp. Therm. Fluid Sci.* **2012**, *42*, 196–203. [CrossRef]
31. Beggs, D.H.; Brill, J.P. An experimental study of two-phase flow in inclined pipes. *J. Pet. Technol.* **1973**, *25*, 607–617. [CrossRef]

Article

Numerical Analysis of Ultrasonic Nebulizer for Onset Amplitude of Vibration with Atomization Experimental Results

Yu-Lin Song [1,2,*], Chih-Hsiao Cheng [3] and Manoj Kumar Reddy [2]

1 Department of Bioinformatics and Medical Engineering, Asia University, Taichung 413, Taiwan
2 Department of Computer Science and Information Engineering, Asia University, Taichung 413, Taiwan; manoj14a0kumar@gmail.com
3 Zeng Hsing Industrial Co., Taichung 411, Taiwan; forcea9931@gmail.com
* Correspondence: d87222007@ntu.edu.tw

Citation: Song, Y.-L.; Cheng, C.-H.; Reddy, M.K. Numerical Analysis of Ultrasonic Nebulizer for Onset Amplitude of Vibration with Atomization Experimental Results. *Water* **2021**, *13*, 1972. https://doi.org/10.3390/w13141972

Academic Editors: Maksim Pakhomov and Pavel Lobanov

Received: 4 June 2021
Accepted: 17 July 2021
Published: 19 July 2021

Abstract: In this study, the onset amplitude of the initial capillary surface wave for ultrasonic atomization of fluids has been implemented. The design and characterization of 485 kHz microfabricated silicon-based ultrasonic nozzles are presented for the concept of economic energy development. Each nozzle is composed of a silicon resonator and a piezoelectric drive section consisting of three Fourier horns. The required minimum energy to atomize liquid droplets is verified by COMSOL Multiphysics simulation software to clarify experimental data. The simulation study reports a minimum vibrational amplitude (onset) of 0.365 μm at the device bottom under the designated frequency of 485 kHz. The experimental study agrees well with the suggested frequency and the amplitude concerning the corresponding surface vibrational velocity in simulation. While operating, the deionized water was initially atomized into microdroplets at the given electrode voltage of 5.96 V. Microdroplets are steadily and continuously formed after the liquid feeding rate is optimized. This newly designed ultrasonic atomizer facilitates the development of capillary surface wave resonance at a designated frequency. A required vibrational amplitude and finite electric driving voltage promote not only the modern development in the green energy industry, but also the exploration of noninvasive, microencapsulated drug delivery and local spray needs.

Keywords: capillary waves; surface wave; subharmonic; resonance; COMSOL; ultrasonic atomizer

1. Introduction

In 1962, Dr. Robert Lang reported that atomization of fluids was caused by a capillary wave formed in an unstable state and small droplets were formed by a collapsing of unstable surface waves. Dr. Robert Lang's work proves the correlation between his atomized droplet sizes relative to Rayleigh's liquid wavelength [1]. Ultrasonic atomized nozzles use high-frequency vibration produced by piezoelectric crystal driving transducers acting upon the nozzle tip that will create capillary waves in a liquid film. Once the amplitude of the capillary surface waves reaches a critical point (due to the power supplied by the generator), surface waves become too tall to support each other and tiny droplets fall off the tip of each wave, resulting in atomization. The basic performances of such devices have been studied experimentally by Lang (1962), Griesshammer and Lierke (1967), Stamm and Pohlman (1965), Sindayihebura et al. (1997), and Topp (1973). The factors that influence the initial droplet size produced are surface tension, frequency of vibration, and viscosity of the liquid. Beyond the range of human hearing, the smallest drop size produces the highest frequencies. Ultrasonic nozzles were first commercialized by Dr. Harvey L. Berger in "Fuel burner with improved ultrasonic atomizer", published 21 January 1975, USA 3861852.

The ultrasonic atomizer is a useful technology and has many practical applications. The ultrasonic atomizer has been successfully applied in areas of biomedical engineering, such as in food industries, chemical coatings, and miniature carriers in pharmaceuticals, as well as for the fabrication of printed electronics and sensors [2,3]. Propagations of ultrasonic surface waves were of great interest in the middle of the 20th century, as they have enabled atomization of incompressible frictionless fluid and have often been known in compact, atomized liquid droplets [4,5]. The control of the atomized droplet process and the characterization of the vaporized droplet size are important issues in obtaining good physical properties and repeatability [6,7]. Moreover, biomedical engineering has successfully applied the technique in drug delivery, pharmacy, and gene therapy [8,9]. Several acoustic microscopy applications were extended to the nondestructive testing of a small structure, intravascular imaging, and ultrasound contrast agent imaging [10,11].

This work develops a practical device for ultrasonic atomization; we explain an onset vibrational amplitude under a designated frequency at the device's bottom [12]. We also aim to know more about the initial surface condition when the energy of ultrasonic atomization of fluids meets the minimum requirement to make a potential green device [13]. The numerical analysis and atomization experimental progress confirmed the theoretical concept to yield tiny atomized droplets under a certain vibrational stimulus [14]. However, the theoretical background was conducted in a steady state. The formation of atomized droplets should be a steadfast and continuous ultrasonic atomization progress. It is necessary to understand more about the dynamic behavior in ultrasonic atomization for real applications [15]. In this study, a dedicated software COMSOL Multiphysics 5.4 is utilized to understand the formation of atomized droplets by the computational fluid dynamics method. The velocity field distribution is solved by finite-element analysis. With regard to the Faraday instability wave, if the effect of the lateral boundary is ignored, the fluctuations generate a standing wave at the surface because of vibration at the bottom. This supports the critical vibrational amplitude (i.e., the characteristic onset of vibrational amplitude) and indicates the constitutive atomized droplets. Further, as the prediction results, we represent the fabrication of the Fourier horn nozzle in three sets at a 485 kHz designated frequency that produces small atomized droplets around the nozzle tip. The feeding liquid flow rate and its vibrational amplitude were carefully measured and statically calculated to understand the manufacturing reliability.

2. Governing Equation

Examinations of surface wave amplification began from the view of Faraday instability in 1831 [16]. Exactly when a vertical vibration is applied for some liquid water on a planar substrate, surface waves are surrounded on a stationary water surface. The destabilized even water surface is accumulated by gatherings of nonlinear standing waves. It is noticed that the periodicity of the working vibration of the capillary surface wave is doubled. In 1871, Kelvin proposed the outstanding condition of capillary surface wave amplification [17]:

$$\lambda = [2\pi T/(\rho f_s^2)]^{1/3} \tag{1}$$

where λ is the wavelength of the capillary waves, T is the surface tension, ρ is the liquid density, and f_s is the frequency of the capillary surface waves.

Rayleigh changed Equation (1) in 1883 as follows:

$$\lambda = [8\pi T/(\rho f^2)]^{1/3} \tag{2}$$

where f is the forcing sound frequency that is observationally two times f_s from practical results.

Benjamin used Mathieu's condition to build a stability diagram for an ideally incompressible frictionless liquid [18]. If a characteristic number and an amplitude of vibration (p, q) lie in any concealed area of the stability diagram as shown in Figure 1, Mathieu's equation would lead to the unstable region that gives rise to the conspicuous standing wave on the fluid surface. If (p, q) situates in one of the unshaded areas, the solution is in a stable district, which implies that the liquid fluid surface is thoroughly quiet without consistent periodic waves.

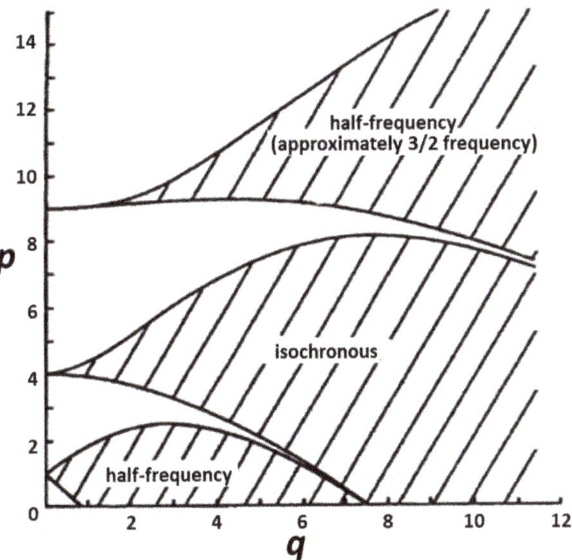

Figure 1. Stability chart for the solutions of Mathieu's equation. Reprinted from Benjamin and Ursell [18].

Eisenmenger then built up a technique for estimating a surface viscosity by methods for rheolinear oscillations [19]. Mathieu equation was conducted by the variable transformation. The minimum capillary wave amplitude is

$$h_{cr} = 2\nu\,[\rho/(T\omega_s)]^{1/3} \tag{3}$$

where $\omega_s = (T\,k^3/\rho + gk)^{\frac{1}{2}}$, ν is the kinematic viscosity of the fluid, k is the capillary wave number, and g is the gravity.

Under low surface viscosity of watery arrangements and feeble vessel boundary conditions, the deliberate capillary surface wave amplitude is half the excitation amplitude with $\pi/4$ phase delay [20]. The experimental approach showed a few frequency vibrations in the range from 2 kHz to 800 kHz. The size of the atomized droplet and the capillary wavelength at a given frequency is proposed as

$$D = 0.34\lambda \tag{4}$$

where D is the diameter of the atomized droplet.

To investigate the formation of atomized droplets, Prof. Perron observed fluid atomization and found that uniform water droplets are formed under a sufficient oscillatory vibration [21]. Prof. Sindayihebura utilized a high-velocity camera technique at high magnification to identify the atomization behavior at the working frequencies of 32 and 50 kHz [22]. The ultrasonic atomized droplet size and velocity distribution of a methanol and glycerol solution were theoretically studied. The ratio between the mean atomized droplet diameter and the vibration wavelength is usually at 0.34–0.36. Prof. Kumar noted that the preferred wave number in the Faraday instability is primarily determined through a Rayleigh–Taylor instability in the deep-water limit (wavelength << liquid depth) by performing linear stability calculations [23].

Ultrasonic atomization studies were for the most part centered on the use of low frequency. For the ultrasonic atomization at a higher frequency, the base required working amplitude remains a provoking topic for researchers investigating surfactant-covered liquid and developing energy-saving devices [24,25].

3. Finite-Element Analysis

COMSOL Multiphysics 5.4 is an advanced software of limited component investigation. It was utilized to demonstrate the basic vibrational amplitude of the silicon-based three-horn Fourier nozzle and the velocity of the surface wave before tests in the following area. A turning hardware model and a laminar stream highlight are referenced for numerical reproductions of surface wave amplification. Navier–Stokes condition is the default mathematic resolver, and the nonlinear impact of liquid fluid is considered to direct arrangements. The vibrational frequency of the silicon-based three-horn ultrasonic nozzle was characterized at f = 485 kHz in the base, at which the characteristic onset amplitude of the capillary surface wave (i.e., the required minimum energy) is calculated for an ultrasonic atomizer. The acquired data support the geometry structure of nozzle detail. The geometry in this reproduction has been characterized in a past investigation [26]. Our endeavor here is to recognize the base plentifulness of atomization under a longitudinal sinusoidal vibration from a nozzle bottom. The investigation case in this report is fluid deionized water, which is broadly useful in numerous applications.

To clarify the onset amplitude of surface waves under a longitudinal sinusoidal oscillation in a liquid-filled nozzle, a moving wall is the setting for generating waves at the nozzle bottom. The velocity in the longitudinal direction is $h_e \times \omega \times \sin(\omega t)$.

The momentum equation is [27]:

$$\rho\frac{\partial u}{\partial t} + \rho(u\cdot\nabla)u = \nabla\cdot\left[-PI + \mu\left(\nabla u + (\nabla u)^T\right) - \frac{2}{3}\mu(\nabla\cdot u)I\right] - \rho\left(g - h_e\omega^2 sin\omega t\right) \quad (5)$$

where u is the velocity field, ρ = 1000 kg/m^3 is the homogeneous density, $\omega = 2\pi f$ is the angular frequency, μ = the dynamic viscosity, h_e = the given amplitude, P = the equilibrium stress at the free fluid surface, and I = the 3×3 identity matrix.

There are at least six components inside a vibrating wavelength. Free surface limit is the default with the goal that the velocity of the liquid fluid surface can be removed from a limited distinction estimate technique at the middle region. The free surface at the fluid, fluid surface, was determined as follows [28]:

$$\left[-PI + \mu\left(\nabla u + (\nabla u)^T\right) - \frac{2}{3}\mu(\nabla\cdot u)I\right]\cdot n = -P_a + T(\nabla_t\cdot n)n - \nabla_t T \quad (6)$$

where P_a is the atmospheric pressure of 1 atm. The full theoretical conduction can be found in the supplementary material.

Because $\frac{\partial\rho}{\partial t} + \nabla\cdot(\rho u) = 0$ and $\nabla\cdot u = 0$ for the zero consumption of mass, Equation (5) can be rewritten as

$$\rho\frac{\partial u}{\partial t} + \rho(u\cdot\nabla)u = -P_a + T(\nabla_t\cdot n)n - \nabla_t T - \rho\left(g - h_e\omega^2 sin\omega t\right) \quad (7)$$

The velocity is expanded with an increased number of periods. In the inset in Figure 2, the schematic outline shows the vibrational amplitude of 0.365 µm at the nozzle tip as demonstrated. In the interim, the velocity amplifies on the droplet surface as set apart by red shading [29].

To estimate the onset vibrational amplitude, a scope of vibrational frequency f of the silicon-based three-horn ultrasonic nozzle was filtered by COMSOL. Figure 2 represents how the capillary surface wave is upgraded inside 400 periods under an onset vibrational amplitude at 0.365 µm. The minor preformed droplets leave the temperamental surface [30] at the most extreme velocity of 0.06 m/s. The vibrational amplitude is upgraded at the center of the Fourier horns by the given vibrational amplitude value.

In this case, based on the water, we use water parameters of density ρ = 1000 (kg/m^3), kinematic viscosity coefficient ν = 1.02 × 10^{-6} (m^2/s), and surface tension T = 7.2 × 10^{-2} (N/m). The frequency range was plotted for the perception of harmonic frequency and the sub-harmonic frequency at 485 and 242.5 kHz individually (Figure 3). It shows that the beginning vibrational amplitude occurred at the assigned frequency of 485 kHz. Further, the wavelength of the capillary surface wave is 19.4 μm, as determined from Figure 4 by the peak-to-peak separation. It is fascinating that the simulated value of 19.4 μm and the theoretical value of 19.8 μm (i.e., given Equation (2)) for the wavelength of the capillary wave are comparable [31].

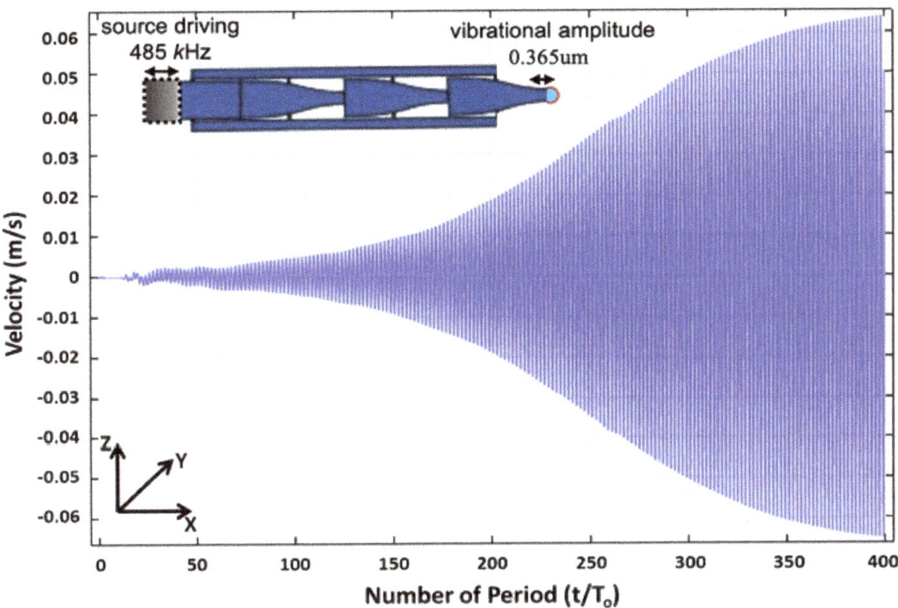

Figure 2. The modeling plot of the velocity in the Z-axis versus time at the nozzle tip of the silicon-based three-horn.

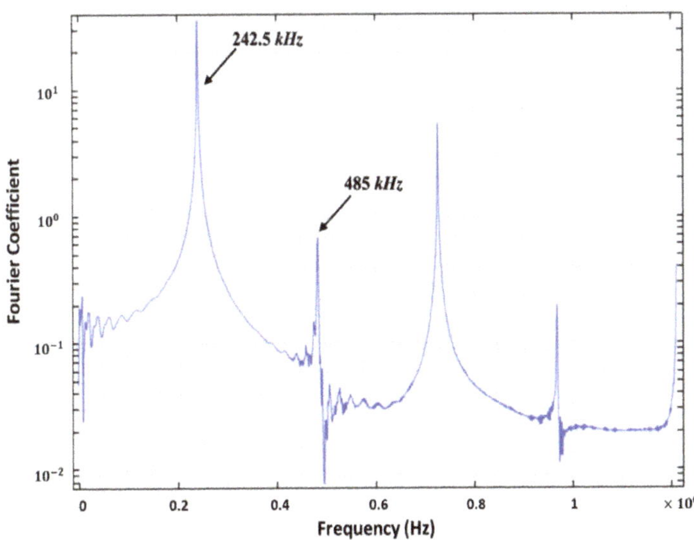

Figure 3. The calculated frequency spectrum at the tip of the silicon-based three-horn. The designated harmonic frequency from COMSOL Multiphysics 5.4 is 485 kHz. As expected, 242.5 kHz is the corresponding subharmonic frequency.

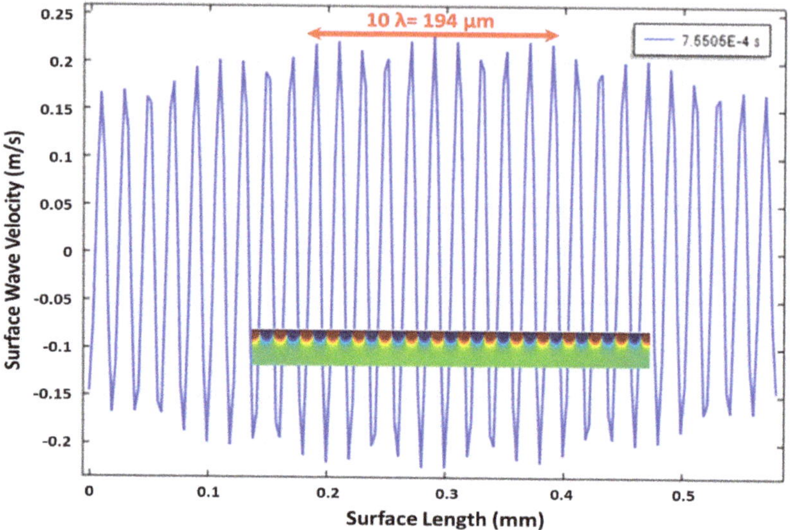

Figure 4. The vibrational velocity is propagated throughout the water surface on the tip of the silicon-based three-horn in the modeling. Inset: the 2D velocity profile at the corresponding position at 7.5505×10^{-4} s simulation time. The red and blue colors present the displacement in the opposite directions.

The lateral velocity profile along Z-axis can be found in the inset of Figure 4 covering the positional velocity of the capillary surface wave. In the supporting material for vibrational velocity (S1) and vibrational amplitude (S2), it can be seen that the vibrational amplitude working at 0.40 μm has a critical upgrade of capillary surface wave propagation. In the supporting material S3, the uprooting and discharge of supporting fluid deionized water along the X-axis are clearly seen. The waves began from PZT plates in the base areas and were collected. Until the beginning vibrational amplitude of 0.365 μm was met, the necessary surface pressure was discharged to atomize modest fluid droplets. Interestingly, beneath the beginning amplitude of 0.365 μm, the capillary surface wave cannot be energized at 0.36, μm as can be found in the supporting material for vibrational velocity (S4) and vibrational amplitude (S5).

4. Experiments

The resonant frequency and its onset amplitude of the capillary surface wave were validated here to help the modeling results. Exploiting the microelectromechanical system (i.e., MEMS), the silicon-based three-horn ultrasonic nozzle was created as delineated in Figure 5a with a focal channel for liquid flow [32]. Three separate Fourier amplifiers were connected in series into one. The magnification obtained by this amplifier can be described in three sections to obtain 8 times magnification. The displacement of the three sections at 500 kHz is in longitudinal vibrational mode. The detailed system and preparation method for the MEMS-based Fourier horn nozzle are described below, including the base areas where PZT plates are to be bounded. The lengths of the silicon-based three-horn and the PZT plates in the transducer drive area were changed by acquiring an ideal high full frequency [33]. The transducer driving square and the resonator square were stuck together to frame a nozzle with a focal channel for liquid trespassing. As deionized water enters the focal channel of the nozzle in 200 μm × 200 μm, a curved thin film remained at the nozzle tip that vibrated at the reverberation frequency of 485 kHz, bringing about the development of standing capillary waves. At the point when the accumulation of the standing capillary waves exceeds a threshold of fluidic surface strain, a shower of droplets/mist is normally created. The normal droplet size from the atomizer was assessed as 7 μm by laser diffraction procedure [34].

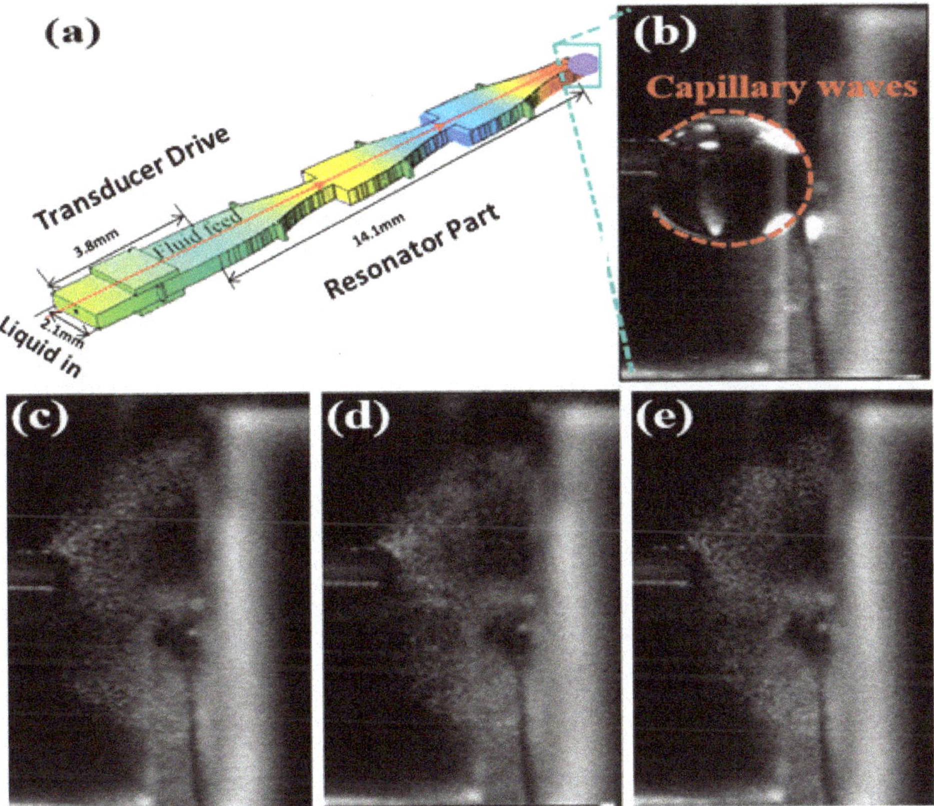

Figure 5. (**a**) Schematic presentation of a silicon-based three-horn ultrasonic nozzle. Optical images obtained from an atomization experiment after voltage was applied for (**b**) 6.49 s, (**c**) 6.50 s (**d**) 43.91 s, and (**e**) 63.59 s. The operating frequency was 484.5 kHz and the applied voltage was 7.9 V under 10 μL/min liquid flow.

Every silicon-based resonator is associated with an arrangement with half-wavelength of ultrasound. The cross-sectional region of the Fourier horn changes bit by bit from the channel to the outlet zone. At the point when the PZT transducer drives a sinusoidal swaying through the Fourier horns, a standing wave is created by the longitudinal vibration of the nozzle tip. Because of energy protection, the vibrational amplitude of the longitudinal direction is collected [35]. Longitudinal vibration was dictated by the Doppler frequency shift in a test estimation of the vibrational frequency and sufficiency at the nozzle tip. Considering working voltages that are changed to mechanical energies, the vibrational amplitude of the three-horn nozzle increased by 2^3 times. As the greatest longitudinal vibration breaks the surface pressure of the liquid, a temperamental fluid capillary surface wave, in the long run, transforms into modest atomized droplets. The progress of the atomized droplets can be seen in Figure 5b–e.

Deionized water was selected as the applied fluid for its adaptable dissolution of different synthetic chemicals. After the underlying atomization of deionized water, the ultrasonic atomized droplets were consistently shaped under a specific sustaining rate, as can be found in the supporting material S6. Capillary surface waves were separated at the droplet of deionized water. The three-horn ultrasonic nozzle was operated at 484.5 kHz with the applied voltage Vp between 6.5 and 7.9 V.

Three gatherings of silicon wafers were examined for the sufficiency check of the longitudinal vibration frequency. Batches were named groups A, B, and C. Arabic numerals address the conveyed Fourier horn nozzle on a comparable silicon wafer. The wafer size was 4 inches in diameter and 530 μm in thickness. The speculative atomization (i.e., longitudinal vibration) was at 485 ± 1 kHz [26]. The cover design was made for the theoretical atomization reason. The sensible results are listed in Table 1. A schematic chart of the arrangement for estimating the longitudinal vibration at the spout tip appears in Figure 5. The pair of PZTs of the silicon-based, ultrasonic spout is driven by the exchanging flow (AC) electrical signal from an Agilent Function Generator Model 33120 A after enhancement by Amplifier Research Model 75 A250 (Amplifier Exploration, Souderton, PA, USA). An Agilent 2-channel Oscilloscope 54621A screens the AC drive signal and gives an outside setting off of the Polytec Ultrasonics Vibrometer Controller Model OFV2700 (Polytech GmbH, Waldbronn, Germany). The regulator is essential for the Polytec laser Doppler vibrometer (LDV) that additionally contains a Polytec Fiber Interferometer Model OFV 511.

Table 1. Comparison of nozzle and atomization parameters. Vp shown is based on the best conditional atomization. Liquid flow rates were carefully adjusted to determine the potential operating conditions.

Nozzle Number	A1	A2	A3	B1	B2	B3	C1	C2	C3	C4	C5
Atomization Frequency (kHz)	484.5	486.5	485	486	484.5	485	488	489	489	489	489
Vp (V)	7.9	6.5	8.7	7.9	6.9	6.5	7.1	6.5	6.6	6.5	6.8
Liquid Flow Rate (μL/min)	10–20	10–30	10–230	10–30	10–20	10–20	10–30	10–100	10–100	10–100	10–50

The test bar in a solitary fiber optic link is engaged and coordinated to the vibrating spout tip surface at a typical frequency. The photograph locator estimates the time subordinate force of the blended light of the test and the reference emissions. The subsequent (beat) recurrence of the blended light is only the Doppler recurrence shift that is relative to the tip vibration speed along with the pivot of the test shaft. The atomization frequency as the nozzle rotates is deliberately estimated by the Polytec LDV at driving voltage Vp before the atomized droplets are consistently and tenaciously encircled at the nozzle head. Note that each reverberating frequency of the instruments is recorded to the extent of 484.5–489 kHz. The differentiation in frequency is an immediate consequence of the holding deviation between any two silicon wafers set up together as resonators, as for a comparative wafer, and slight misalignment in the collecting methodology [31]. The liquid stream rate is also affected by collection error. A higher liquid stream rate (~100 μL/min) was tested and suffered due to MEMS nozzle fabrication.

A schematic diagram for measuring the longitudinal vibration at the tip of the nozzle is shown in Figure 6. The pair of PZTs of the silicon-based ultrasonic nozzle are driven by the AC electrical signal from an Agilent Function Generator 33120 A after amplification by Amplifier Research (AR) Model 75A250 (Amplifier Research, Souderton, PA, USA). A 2-channel 54621A oscilloscope monitors the AC drive signal and provides an external triggering of the Polytec Vibrometer Controller Model OFV2700 (Polytech GmbH, Waldbronn, Germany) [26]. The controller is part of the Polytec laser Doppler vibrometer (LDV) that also contains a Polytec Fiber Interferometer Model OFV 511. A He-Ne laser at 632.8 nm wavelength in the interferometer is divided into a reference beam and a probe beam. The probe beam traveling in a single fiber optic cable is focused and directed to the vibrating nozzle tip surface at normal incidence. The photodetector measures the time-dependent intensity of the mixed light of the probe and the reference beams. The resulting (beat) frequency of the mixed light is just the Doppler frequency shift that is proportional to the tip vibration velocity along the axis of the probe beam [36]. Longitudinal vibration at the nozzle base also was measured to provide a reference for experimental determination of the vibration amplitude magnification (or gain) at the nozzle tip.

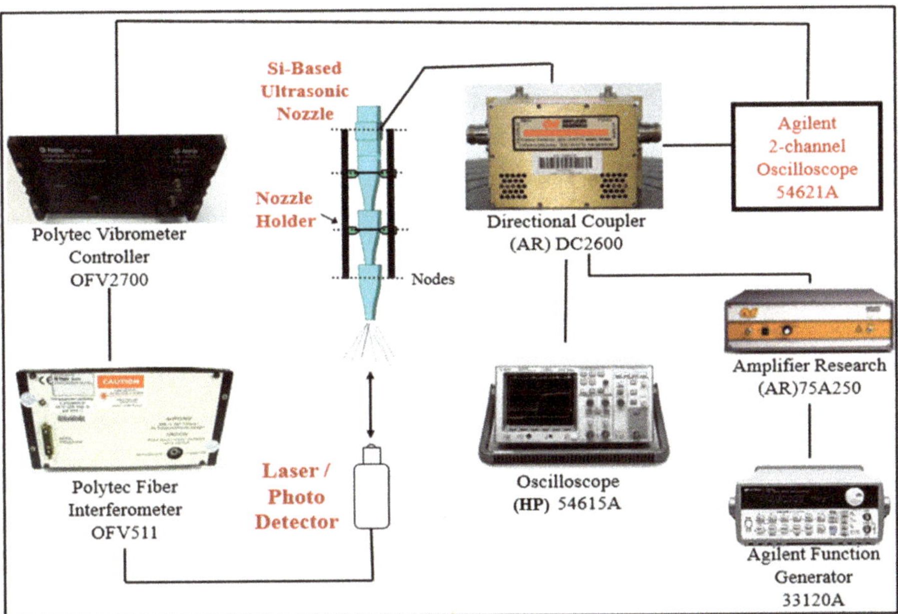

Figure 6. Schematic diagram for measuring the longitudinal vibration at the nozzle tip [26].

In Figure 7, the results are depicted in blue color. The calibration line is a function of the electrode voltage. The red coordinate (5.96, 9.125) on the calibration line is the suggested onset vibrational amplitude of 0.365 µm as the critical longitudinal displacement of capillary waves from modeling. The red coordinate (6.43, 10) confirms the linear regression of the applied electrode voltage for the Polytec LDV output signal in experiments. Note that the Polytec LDV output signal 1 V = 0.04 µm longitudinal displacement.

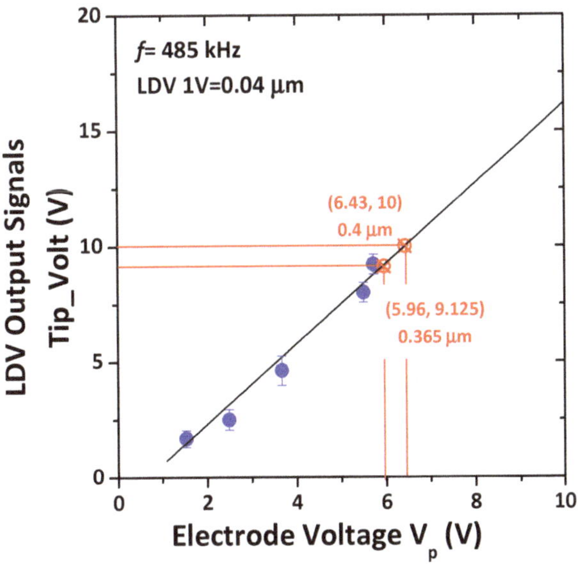

Figure 7. Output signals of longitudinal vibration at the nozzle tip versus applied electrode voltage at the base of the silicon-based three-horn Fourier nozzle.

In Figure 7, the yield signal from the Polytec LDV estimations of the longitudinal vibrations of the three-horn ultrasonic nozzle at various driving voltages is shown. The straight backslide shows that the vibrational plenteousness of the response variable can be constrained by the working voltage. The developed change line predicts the longitudinal evacuating of the silicon-based three-horn Fourier nozzle as shown by the applied electrode voltage, similarly to the vibrational sufficiency. The exploratory result gives affirmations of the starting adequacy of capillary surface wave in high kHz for ultrasonic atomization. It can be noticed that the base applied electrode voltage of 5.96 V is adequate to produce strong capillary surface waves that can atomize deionized water at the nozzle tip. The details were elucidated in the modeling. The required minimum energy to atomize liquid drops is verified, and the modeling results correspond well to the experimental observations.

5. Conclusions

This study of ultrasonic atomization of fluids for onset amplitude of initial capillary surface wave phenomena reports simulation results along with experimental verification. COMSOL Multiphysics 5.4 modeling predicts the threshold vibrational amplitude of 0.365 μm at 485 kHz. The constraint of the surface tension breaks when the amplitude of the capillary surface wave is increased as time elapses. Based on the basic physics that describes the water droplet formation process, the atomization of fluid through ultrasound waves in a piezoelectrically actuated nozzle was performed to determine the critical surface velocity of atomized droplets. Experimental studies prove that the surface plane with standing capillary surface waves would produce microdroplets at a designated electrode voltage of 5.96 V. The output voltage of longitudinal vibration at the nozzle bottom scales the driving voltage, and the frequency of microdroplets scales with capillary time, showing that silicon-based three Fourier-horn ultrasonic nozzles conform to the design concept from simulations. The advanced design of high-frequency ultrasonic nozzles demonstrated the minimum amplitude requirement of longitudinal vibration for the nozzle tip. The silicon-based three-horn Fourier nozzle is not only compact but also an energy-saving device. It promotes the further development of single-frequency ultrasonic nebulizers, especially in the green energy industry.

Supplementary Materials: The following are available online at https://www.mdpi.com/article/10.3390/w13141972/s1, Video S1: Velocity-0.40um.avi, Video S2: Amplitude-0.40um.avi, Video S3: 500kHz-Displacement.avi, Video S4: Velocity-0.36um.avi, Video S5: Amplitude-0.36um.avi, Video S6: B1200_3H3.avi.

Author Contributions: Conceptualization, Y.-L.S.; investigation, Y.-L.S. and C.-H.C.; methodology, Y.-L.S. and C.-H.C.; validation, C.-H.C.; writing—review and editing, C.-H.C. and Y.-L.S.; software, M.K.R.; data curation, M.K.R.; writing—original draft, M.K.R.; funding acquisition, Y.-L.S. All authors have read and agreed to the published version of the manuscript.

Funding: Ministry of Science and Technology, Taiwan, through grant MOST 106-2221-E-468-015 and MOST 110-2622-E-468-002.

Institutional Review Board Statement: Not applicable.

Informed Consent Statement: Not applicable.

Data Availability Statement: The study did not report any data.

Acknowledgments: This work was supported in part by the Ministry of Science and Technology, Taiwan, through grant MOST 106-2221-E-468-015.

Conflicts of Interest: The authors declare no conflict of interest.

References

1. Lang, R.J. Ultrasonic Atomization of Liquids. *J. Acoust. Soc. Am.* **1962**, *34*, 6–8. [CrossRef]
2. Schehl, N.; Kramb, V.; Dierken, J.; Aldrin, J.; Schwalbach, E.; John, R. Ultrasonic assessment of additive manufactured Ti-6Al-4V. *AIP Conf. Proc.* **2018**, *1949*, 020008.
3. Taheri, H.; Koester, L.; Bigelow, T.; Bond, L.J. Finite element simulation and experimental verification of ultrasonic non-destructive inspection of defects in additively manufactured materials. *AIP Conf. Proc.* **2018**, *1949*, 020011.
4. Tan, M.K.; Friend, J.R.; Matar, O.K.; Yeo, L.Y. Capillary wave motion excited by high frequency surface acoustic waves. *Phys. Fluids* **2010**, *22*, 112112. [CrossRef]
5. Janjic, J.; Kruizinga, P.; Van Der Meulen, P.; Springeling, G.; Mastik, F.; Leus, G.; Bosch, J.G.; van der Steen, A.F.; van Soest, G. Structured ultrasound microscopy. *Appl. Phys. Lett.* **2018**, *112*, 251901. [CrossRef]
6. Wei, L.-J.; Oxley, C.H. Carbon based resistive strain gauge sensor fabricated on titanium using micro-dispensing direct write technology. *Sens. Actuators A Phys.* **2016**, *247*, 389–392. [CrossRef]
7. White, R.M.; Voltmer, F.W. Ultrasonic Surface-wave amplification in cadmium sulfide. *Appl. Phys. Lett.* **1966**, *8*, 40–42. [CrossRef]
8. Yang, H.-W.; Hua, M.-Y.; Hwang, T.-L.; Lin, K.-J.; Huang, C.-Y.; Tsai, R.-Y.; Ma, C.-C.M.; Hsu, P.-H.; Wey, S.-P.; Hsu, P.-W.; et al. Non-Invasive synergistic treatment of brain tumors by targeted chemotherapeutic delivery and amplified focused ultrasound-hyperthermia using magnetic nanographene oxide. *Adv. Mater.* **2013**, *25*, 3605–3611. [CrossRef]
9. Yang, F.-Y.; Chang, W.-Y.; Lin, W.-T.; Hwang, J.-J.; Chien, Y.-C.; Wang, H.-E.; Tsai, M.-L. Focused ultrasound enhanced molecular imaging and gene therapy for multifusion reporter gene in glioma-bearing rat model. *Oncotarget* **2015**, *6*, 36260–36268. [CrossRef]
10. Katchadjian, P.; García, A.; Brizuela, J.; Camacho, J.; Álvarez-Arenas, T.G.; Chiné, B.; Mussi, V. Ultrasonic techniques for the detection of discontinuities in aluminum foams. *AIP Conf. Proc.* **2017**, *1806*, 090018.
11. Peruzzini, D.; Viti, J.; Tortoli, P.; Verweij, M.D.; de Jong, N.; Vos, H.J. Ultrasound Contrast Agent Imaging: Real-Time Imaging of the Superharmonics. *AIP Conf. Proc.* **2015**, *1685*, 040011.
12. Saffari, N. Therapeutic Ultrasound—Exciting Applications and Future Challenges. *AIP Conf. Proc.* **2018**, *1949*, 020001.
13. Breitenbach, J.; Kissing, J.; Roisman, I.V.; Tropea, C. Characterization of secondary droplets during thermal atomization regime. *Exp. Therm. Flu. Sci.* **2018**, *98*, 516–522. [CrossRef]
14. Song, Y.L.; Cheng, C.H.; Chang, L.M.; Lee, C.F.; Chou, Y.F. Novel device to measure critical point "Onset" of capillary wave and interpretation of Faraday instability wave by numerical analysis. *Ultrasonics* **2012**, *52*, 54–61. [CrossRef]
15. Deepu, P.; Peng, C.; Moghaddam, S. Dynamics of ultrasonic atomization of droplets. *Exp. Therm. Flu. Sci.* **2018**, *92*, 243–247. [CrossRef]
16. Faraday, M. On a peculiar class of acoustical figures; and on certain forms assumed by groups of particles upon vibrating elastic surfaces. In *Abstracts of the Papers Printed in the Philosophical Transactions of the Royal Society of London;* The Royal Society: London, UK, 1837.
17. Barreras, F.; Amaveda, H.; Lozano, A. Transient high-frequency ultrasonic water atomization. *Exp. Fluids* **2002**, *33*, 405–413. [CrossRef]
18. Benjamin, T.B.; Ursell, F. The stability of the plane free surface of a liquid in vertical periodic motion. *Proc. R. Soc. Lond. Ser. A* **1954**, *225*, 505–515.
19. Eisenmenger, W. Dynamic properties of the surface tension of water and aqueous solutions of surface-active agents with standing capillary waves in the frequency range from 10 kc/s to 1.5 Mc/s. *Acustica* **1959**, *9*, 327.
20. Tsai, C.S.; Mao, R.W.; Lin, S.K.; Zhu, Y.; Tsai, S.C. Faraday instability-based micro droplet ejection for inhalation drug delivery. *Technology* **2014**, *2*, 75–81. [CrossRef]
21. Perron, R.R. The design and application of a reliable ultrasonic atomizer. *IEEE Trans. Sonics Ultrason.* **1967**, *14*, 149–152. [CrossRef]
22. Sindayihebura, D.; Dobre, M.; Bolle, L. Experimental study of thin liquid film ultrasonic atomization. In Proceedings of the 4th World Conference on Experimental Heat Transfer, Fluid Mechanics and Thermodynamics 1997, Brussels, Belgium, 2–6 June 1997.
23. Kumar, S. Mechanism for the Faraday Instability in Viscous Liquids. *Phys. Rev. E* **2000**, *62*, 1416–1419. [CrossRef] [PubMed]
24. Kumar, S.; Matar, O.K. On the Faraday instability in a surfactant-covered liquid. *Phys. Fluids* **2004**, *16*, 39–46. [CrossRef]
25. Tsai, S.C.; Tsai, C.S. Linear theory on temporal instability of megahertz faraday waves for monodisperse microdroplet ejection. *IEEE Trans. Ultrason. Ferroelectr. Freq. Control* **2013**, *60*, 1746–1755. [CrossRef] [PubMed]
26. Tsai, S.C.; Song, Y.L.; Tseng, T.K.; Chou, Y.F.; Chen, W.J.; Tsai, C.S. High-frequency, silicon-based ultrasonic nozzles using multiple fourier horns. *IEEE Trans. Ultrason. Ferroelectr. Freq. Control* **2004**, *51*, 277–285. [CrossRef]
27. Khaire, R.A.; Gogate, P.R. Novel approaches based on ultrasound for spray drying of food and bioactive compounds. *Dry. Technol.* **2020**, 1–22. [CrossRef]
28. Altay, R.; Sadaghiani, A.K.; Sevgen, M.I.; Şişman, A.; Koşar, A. Numerical and experimental studies on the effect of surface roughness and ultrasonic frequency on bubble dynamics in acoustic cavitation. *Energies* **2020**, *13*, 1126. [CrossRef]
29. Tsai, C.S.; Mao, R.W.; Tsai, S.C.; Shahverdi, K.; Zhu, Y.; Lin, S.K.; Hsu, Y.-H.; Boss, G.; Brenner, M.; Mahon, S.; et al. Faraday Waves-Based Integrated Ultrasonic Micro-Droplet Generator and Applications. *Micromachines* **2017**, *8*, 56. [CrossRef] [PubMed]
30. Tsai, S.C.; Song, Y.L.; Tsai, C.S.; Chou, Y.F.; Cheng, C.H. Ultrasonic atomization using MHz silicon-based multiple-Fourier horn nozzles. *Appl. Phys. Lett.* **2006**, *88*, 014102. [CrossRef]
31. Song, Y.L.; Bandi, L. Design and simulation of the new ultrasonic atomizer using silicon-based with one step resonator. *Results Phys.* **2020**, *18*, 103166. [CrossRef]

32. Li, Z.; Ning, H.; Liu, L.; Xu, C.; Li, Y.; Zeng, Z.; Liu, F.; Hu, N. Fabrication of bagel-like graphene aerogels and its application in pressure sensors. *Smart Mater. Struct.* **2019**, *28*, 055020. [CrossRef]
33. Mishra, H.; Hehn, M.; Lacour, D.; Elmazria, O.; Tiercelin, N.; Mjahed, H.; Dumesnil, K.; Watelot, S.P.; Polewczyk, V.; Talbi, A.; et al. Intrinsic versus shape anisotropy in micro-structured magnetostrictive thin films for magnetic surface acoustic wave sensors. *Smart Mater. Struct.* **2019**, *28*, 12LT01. [CrossRef]
34. Avvaru, B.; Patil, M.N.; Gogate, P.R.; Pandit, A.B. Ultrasonic atomization: Effect of liquid phase properties. *Ultrasonics* **2006**, *44*, 146–158. [CrossRef]
35. Simon, J.C.; Sapozhnikov, O.A.; Khokhlova, V.A.; Crum, L.A.; Bailey, M.R. Ultrasonic atomization of liquids in drop-chain acoustic fountains. *J. Fluid Mech.* **2015**, *766*, 129–146. [CrossRef] [PubMed]
36. Berger, H.L. Fuel Burner with Improved Ultrasonic Atomizer. U.S. Patent 3,861,852, 21 January 1975.

Article

The Effect of Wall Shear Stress on Two Phase Fluctuating Flow of Dusty Fluids by Using Light Hill Technique

Dolat Khan [1], Ata ur Rahman [2], Gohar Ali [2], Poom Kumam [1,3,4,*], Attapol Kaewkhao [5,*] and Ilyas Khan [6]

1 Department of Mathematics, Faculty of Science, King Mongkut's University of Technology Thonburi (KMUTT), 126 Pracha Uthit Rd., Bang Mod, Thung Khru, Bangkok 10140, Thailand; dolat.ddk@gmail.com
2 Department of Mathematics, City University of Science & Information Technology, Peshawar 25000, KPK, Pakistan; ataurrahman.at80@gmail.com (A.u.R.); goharali.cu@gmail.com (G.A.)
3 Center of Excellence in Theoretical and Computational Science (TaCS-CoE), Faculty of Science, King Mongkut's University of Technology Thonburi (KMUTT), 126 Pracha Uthit Rd., Bang Mod, Thung Khru, Bangkok 10140, Thailand
4 Department of Medical Research, China Medical University Hospital, China Medical University, Taichung 40402, Taiwan
5 Data Science Research Center, Department of Mathematics, Faculty of Science, Chiang Mai University, Chiang Mai 50200, Thailand
6 Department of Mathematics, College of Science Al-Zulfi, Majmaah University, Al-Majmaah 11952, Saudi Arabia; i.said@mu.edu.sa
* Correspondence: poom.kumam@mail.kmutt.ac.th (P.K.); attapol.k@cmu.ac.th (A.K.)

Abstract: Due to the importance of wall shear stress effect and dust fluid in daily life fluid problems. This paper aims to discover the influence of wall shear stress on dust fluids of fluctuating flow. The flow is considered between two parallel plates that are non-conducting. Due to the transformation of heat, the fluid flow is generated. We consider every dust particle having spherical uniformly disperse in the base fluid. The perturb solution is obtained by applying the Poincare-Lighthill perturbation technique (PLPT). The fluid velocity and shear stress are discussed for the different parameters like Grashof number, magnetic parameter, radiation parameter, and dusty fluid parameter. Graphical results for fluid and dust particles are plotted through Mathcad-15. The behavior of base fluid and dusty fluid is matching for different embedded parameters.

Keywords: wall shear stress; oscillating two-phase fluctuation flow; heat transfer; magnetohydrodynamic (MHD); dust particles

Citation: Khan, D.; Rahman, A.u.; Ali, G.; Kumam, P.; Kaewkhao, A.; Khan, I. The Effect of Wall Shear Stress on Two Phase Fluctuating Flow of Dusty Fluids by Using Light Hill Technique. *Water* **2021**, *13*, 1587. https://doi.org/10.3390/w13111587

Academic Editors: Maksim Pakhomov and Pavel Lobanov

Received: 22 April 2021
Accepted: 30 May 2021
Published: 4 June 2021

Publisher's Note: MDPI stays neutral with regard to jurisdictional claims in published maps and institutional affiliations.

1. Introduction

Fluid flow inserted with similar non-miscible inert solid particles is admitted as the two-phase structure of fluid. The fields of technology and engineering accommodate various uses of the flow of the gas-particles mixture. Nuclear reactors with gas-solid feeds, cooling of nuclear reactors, solid rocket exhaust nozzles, electrostatic drizzle, ablation cooling, polymer technology, the moving blast waves over the Earth's surface, the distillation of crude oil, environmental poison, the industry of petroleum, fluidized beds, transmit of powdered materials, physiological flows and various other fields of technologies are the practical examples where the dusty fluid are mostly applicable [1,2]. Different types of multiphase flows can be seen in the literature. However, the most familiar type of such flows, which are multiphase is two-phase flows. In these multiphase flows, liquid and liquid flow, gas and liquid flow, liquid and solid flow, solid and gas flow are frequently discussed by researchers [3,4]. In turbulent flows, the liquid droplets vaporization is a crucial multiphase flow area due to an extensive measure of energy used for heating. Electrical and impulse power generation is borrowed against liquid fuels that have been transformed into atomized sprays. The cooperation of the discharged vapors of droplets with the turbulent flow formed immensely complicated problems in such situations. A comprehensive explanation of the fundamental flow processes and droplets involved in these situations can be initiated [5–7].

The researchers are very keen to study the dusty multiphase flows over the last few decades because of their numerous applications. In this regard, various researchers investigate preliminary and theoretical modeling of particle phase viscosity in a multiphase dusty fluid [8–11], but keep in mind Soo [12] is the first who presented the fundamental theory of multiphase flows. Other useful applications consist of dust particles in nuclear processing, in boundary layers contains soil emancipation which is occurred by natural winds, in aerodynamic refusal of plastic sheets, landing vehicle in a cloud designed during a nuclear detonation, and lunar surface extinction by the dust entrainment are investigated by many researchers in the literature [13–16]. Zhou et al. [17] discuss the fluid particles in translational motion and study the converging and diverging behavior in a microchannel. In another paper, Zhau et al. [18] established a model in which he examines the deformable interaction of particles, and he studies the contact of dielectrophoresis (DEP) thoroughly in the presence of an electric field generated by the alternating current. Recently, Ali et al. [19] present an article on fluctuating flow of absorbing heat viscoelastic dusty fluid with free convection and MHD effect past in a horizontal channel. In this reported study, the researcher discussed the consequence of different parameters on fluid velocity and particles. They investigate the solution for the fluid velocity and the dust particle's velocity by applying the Poincare-Light Hill technique. Similarly, in another paper, Ali et al. [20] established the study about the dusty viscoelastic fluid of two-phase fluctuating flow with heat transfer between rigid plates which are non-conducting and examine the connected effect of the heat transfer and magnetic field on the dusty viscoelastic fluid which is conducting electrically with the help of Light-Hill technique.

Furthermore, Attia et al. [21] work out the fallout of different substantial parameters on the steady flow of MHD incompressible non-Newtonian Oldroyd dusty fluid in a circular pipe and also study the consequence of Hall current. In this particular article, the author investigates the characteristics of the particle phase viscosity and non-Newtonian fluid, flow rates in terms of volume, and the coefficient of skin fraction for both the particle phase and fluid. Keeping in mind Ali et al. [22] present an article about second-grade fluid in which they present closed-form solutions of free convection unsteady flow. In this article, the author investigates the influence of different parameters on the velocity profile of second-grade fluid-like Grashof number, Prandtl number, and viscoelastic parameter.

Multiphase flow regimes have significant consideration because of their useful applications [23]. There is two specific access that has been used commonly. According to this particular access, if we study the first way, we can observe that the continuous phase's motion is not significantly affected by the dispersed phase. This phase's motion is so minute. This approach is commonly admitted as the "alter phased approach," which is also familiar with the Lagrangian approach. This approach is applied broadly in situations where the particles deal with dispersed phase like in sprays, atomization, and droplets [23].

On the other hand, the second approach is related to the two phases combined so that each phase precisely manipulates the altitude and motion of the other stage [23]. The second approach is called the method of dense phase. It is also known as the Eulerian approach. This approach is beneficial in solid-gas flows [24], aerial transmitting [25], in fluidization and is defined for a variety of uses and applications in suspensions [26,27].

The Light-Hill method was introduced by M. J. Light-Hill in 1949 [28]. He obtained uniformly reasonable approximate solutions for various classes of partial and ordinary differential equations. After that, this method was used by many researchers successfully in the research field and solve ordinary and partial differential Equations [29]. As far as wall-shear stress is concerned, many researchers present different problems in discussing wall-shear stress and interpret other behaviors of the various parameters on the fluid velocity. In this regard, Grobe et al. [30] study the boundary layer wall shear-stress of the turbulent wind tunnel with a high Reynolds number. In a similar way, Amili et al. [31]

represent wall shear-stress distribution in a turbulent flow in the channel and discuss the different influence of parameters on the fluid's velocity. Keeping in mind, Orlu et al. [32] described the study on the fluctuating wall shear stress in zero pressure-gradient turbulent boundary layers flow. Recently, Mob et al. [33] present a paper in which they describe the study of the impact of wall-shear stress on the evolution and execution of electrochemically alive biofilm. The author referred to some recent contributions in the field [34–37].

In the literature mentioned above, the authors have considered different types of dusty fluids. Some of the dusty fluids are electrically conducting, and some are investigated with heat transfer and energy equations. According to our utmost knowledge, no investigation has been disclosed about the effect of wall shear stress on two phases of the fluctuating flow of dusty fluid by solving the Light-Hill technique. Therefore, we have to examine and study the different behavior of velocity of the problem theoretically and graphically in the present work. Hence, the purpose of this study is to interrogate the flow of fluid ingrained with dust particles along with transfer heat over the bounded plates.

2. Formulation of the Problem

The incompressible, unidirectional, and one-dimensional electrically conducting the unsteady flow of dusty fluid along the x-axis has been considered between two parallel plates. The magnetic field B_0 is transversely applied to the fluid, and due to the small size of the induced magnetic field, the emission is so small; therefore, the electric field inside is ignored. Also, flow origination is induced by heat transfer, the upper plate temperature, and the lower plate temperature T_ω. The wall shear stress is applied to the lower plate, where the upper plate fluctuates with free stream velocity $U(t) = u_0\left(1 + \frac{\varepsilon}{2}\left(e^{i\omega t} + e^{-i\omega t}\right)\right)$. $u(y,t)$ and $v(y,t)$ denoted the velocities of fluid and dust particles, respectively. As shown in Figure 1. The effect of the equation of energy radiation is also getting hold of into description. By apply the assumptions of Bossiness resemblance and in sequence to escape similarities, the equation of momentum and energy is.

$$\frac{\partial u}{\partial t} = \upsilon\frac{\partial^2 u}{\partial y^2} + \frac{K_0 N_0}{\rho}(v - u) - \frac{\sigma B_0^2 u}{\rho} + g\beta_T(T - T_\infty) \tag{1}$$

$$\rho c_p \frac{\partial T}{\partial t} = k\frac{\partial^2 T}{\partial y^2} - \frac{\partial q_r}{\partial y} \tag{2}$$

where $-\frac{\partial q_r}{\partial y} = 4\alpha_0(T - T_\infty)$.

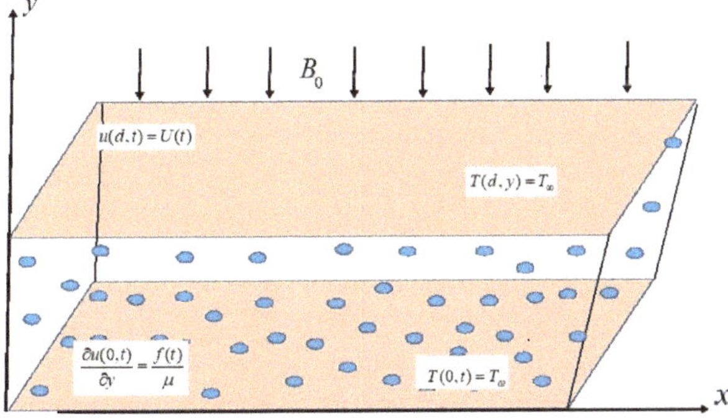

Figure 1. Geometry of the problem.

Equations (1) and (2), which are constantly spreader in the viscoelastic fluid, q_r, υ, c_p, k, g, ρ, α_0, σ, B_0, N_0, β_T and K_0 are show the radiation heat flux, kinematic viscosity, specific heat capacity, thermal conductivity, gravitational acceleration, fluid density, mean radiation absorption coefficient, electrical conductivity, magnetic field, number of density of the dust particle which is supposed to be constant, coefficient of thermal expansion, and stock's resistance coefficient respectively.

This can be represented by Newton's law of motion as:

$$m\frac{\partial v}{\partial t} = K_0(u - v) \tag{3}$$

where m represents the average mass of dust particles.

The physical boundary conditions are:

$$\left.\begin{array}{ll} \frac{\partial u(0,t)}{\partial y} = \frac{f(t)}{\mu} & t > 0, \quad u(d,t) = U(t) \\ T(0,t) = T_\omega, & T(d,t) = T_\infty \end{array}\right\} \tag{4}$$

where $U(t) = u_0\left(1 + \frac{\varepsilon}{2}\left(e^{i\omega t} + e^{-i\omega t}\right)\right)$.

To find the velocity of the dust particle let assume the following solution, obtained through Poincare-Light Hill Technique [28]:

$$v(y,t) = v_0(y)e^{i\omega t} \tag{5}$$

By using Equation (5) in Equation (3) velocity of the dust particles will be.

$$v(y,t) = (\frac{K_0}{mi\omega + K_0})u(y,t) \tag{6}$$

Put Equation (6) 1n Equation (1). So, Equation (1) becomes,

$$\frac{\partial u}{\partial t} = v\frac{\partial^2 u}{\partial y^2} + \frac{K_0 N_0}{\rho}\left\{\left(\frac{K_0}{mi\omega + K_0}\right)u - u\right\} - \frac{\sigma B_0^2 u}{\rho} + g\beta_T(T - T_\infty) \tag{7}$$

By use of the following dimensionless variables.

$$u^* = \frac{u}{u_0}, y^* = \frac{y}{d}, t^* = \frac{u_0 t}{d}, \theta^* = \frac{T - T_\infty}{T_\omega - T_\infty} \tag{8}$$

For simplicity (*) sign has been ignored. So Equation (2), and Equation (7) becomes:

$$\frac{\partial u}{\partial t} = \frac{\partial^2 u}{\partial y^2} - (M + K_1 - K_2)u + Gr\theta \tag{9}$$

$$Pe\frac{\partial \theta}{\partial t} = \frac{\partial^2 \theta}{\partial y^2} + N^2\theta \tag{10}$$

with dimensionless physical conditions are:

$$\left.\begin{array}{l} \frac{\partial u(0,t)}{\partial y} = f(t) \quad , u(1,t) = U(t) = 1 + \frac{\varepsilon}{2}\left(e^{i\omega t} + e^{i\omega t}\right) \\ \theta(0,t) = 1 \quad , \quad \theta(1,t) = 0 \end{array}\right\} \tag{11}$$

where,
$M = \frac{\sigma B_0^2 d^2}{\rho v}, Gr = \frac{g\beta_T d^2(T_\omega - T_\infty)}{u_0 v}, K_2 = \frac{K_0^2 N_0 d^2}{\rho v(mi\omega + K_0)}, K_1 = \frac{K_0 N_0 d^2}{\rho v}, Pe = \frac{\rho c_p u_0 d}{k}, N^2 = \frac{4\alpha_0^2 d^2}{k}$.
where h_s is a heat transfer coefficient, M is a magnetic variable, Gr is Grashof number, K_1 dusty fluid variable, K_2 Dust particles parameter, P_e Peclet number,N radiation variable.

Consider the following assume periodic solutions for energy equation obtained through Poincare-Light Hill Technique in [28];

$$\theta(y,t) = \theta_0(y) + \theta_1(y)e^{i\omega t} \tag{12}$$

By solving energy equation with the help of above assumed periodic solution we get:

$$\theta(y,t) = \frac{\sin(N - Ny)}{\sin(N)} \tag{13}$$

By putting Equation (13) in Equation (9), we get:

$$\frac{\partial u}{\partial t} = \frac{\partial^2 u}{\partial y^2} - (M + K_1 - K_2)u + Gr\left\{\frac{\sin(N - Ny)}{\sin(N)}\right\} \tag{14}$$

3. The Solution to the Problem

Basic idea of Poincare-Light Hill Technique and model framework [28]:
The original model of lighthill is firstly introduced by Lighthill in 1949 [28]

$$(x + \varepsilon y)\frac{dy}{dx} + q(x)y = r(x),\ 0 \le x \le 1 \tag{15}$$

$$y(1) = b > 1, \tag{16}$$

With condition

$$q(0) \ne 0. \tag{17}$$

This last condition is one to which we shall return later. If one tries to find y as a series in ε,

$$y \sim y_0(x) + \varepsilon y_1(x) + \varepsilon^2 y_2(x) + \cdots \tag{18}$$

$$x\frac{dy_0}{dx} + q(x)y_0 = r(x) \tag{19}$$

whose homogeneous part has a regular singular point at the origin. The original equation, as a simple phase plane analysis shows, does not have a singularity at the origin, and thus the perturbation series has a false singularity.

Lighthill's idea was to move the singularity back out of the domain of interest by introducing a new, slightly stretched, independent variable z by the implicit equation

$$x = z + \varepsilon x_1(z) + \varepsilon^2 x_2(z) + \cdots \tag{20}$$

And then to look for a solution y in the form

$$y = y_0(x) + \varepsilon y_1(z) + \varepsilon^2 y_2(z) + \cdots \tag{21}$$

On the light of Equation (21) the following assume periodic solutions for Equation (14) obtained through Poincare-Light Hill Technique in [28] is considered as follow.

$$u(y,t) = F_0(y) + \frac{\varepsilon}{2}\left(F_1(y)e^{i\omega t} + F_2(y)e^{-i\omega t}\right) \tag{22}$$

To find out the results for $F_0(y), F_1(y)$ *, and* $F_2(y)$, incorporate Equation (22) in (14) we get:

$$F_0(y) = \left(1 - \frac{(f(t) + H)\sinh(\sqrt{m_1})}{\sqrt{m_1}.\cosh(\sqrt{m_1})}\right)\cosh(y\sqrt{m_1}) + \left(\frac{f(t) + H}{\sqrt{m_1}}\right)\sinh(y\sqrt{m_1}) + A\left\{\frac{\sin(N - Ny)}{\sin(N)}\right\} \tag{23}$$

where $H = \frac{A \cdot N \cdot \cos(N)}{\sin(N)}$, and $m_1 = M + K_1 - K_2$, $A = \frac{Gr}{m_1}$.

$$F_1(y) = \frac{\cosh(y\sqrt{m_2})}{\cosh(\sqrt{m_2})} \tag{24}$$

where $m_2 = m_1 + i\omega$

$$F_2(y) = \frac{\cosh(y\sqrt{m_3})}{\cosh(\sqrt{m_3})} \tag{25}$$

where $m_3 = m_1 - i\omega$.

In last, we put the results in Equation (22) from Equations (23)–(25), we obtained the following output:

$$u(y,t) = \left\{ \begin{array}{l} \left(1 - \frac{(f(t)+H)\sinh(\sqrt{m_1})}{\sqrt{m_1}.\cosh(\sqrt{m_1})}\right)\cosh(y\sqrt{m_1}) + \left(\frac{f(t)+H}{\sqrt{m_1}}\right)\sinh(y\sqrt{m_1}) \\ +A\left\{\frac{\sin(N-Ny)}{\sin(N)}\right\} + \frac{\varepsilon}{2}\left(\frac{\cosh(y\sqrt{m_2})}{\cosh(\sqrt{m_2})}\right)e^{i\omega t} + \frac{\varepsilon}{2}\left(\frac{\cosh(y\sqrt{m_3})}{\cosh(\sqrt{m_3})}\right)e^{-i\omega t} \end{array} \right\}. \tag{26}$$

4. Graphical Results and Discussion

In this work, the unsteady motion because of an infinite plate that addresses wall-shear stress to a two-phase fluctuating flow of dusty fluids is considered utilizing the light hill technique. The solutions that have been achieved satisfy all the introduced initial and boundary conditions. To find out some specific and vital information about the repercussions of different flow parameters on dusty magnetic particles and fluid velocities. Certain numerical simultaneous results have been made with Mathcad-15 software. These influences of various physical parameters like magnetic parameter M, dusty fluid parameter K_2, Grashof number Gr and radiation variable N are graphically shown in this section. According to these graphs, we get different results for the profile of fluid velocities and the velocities of dusty particles, briefly discussed in the following paragraph. The estimated values of parameters for all figures are taken from Table 1.

Table 1. Description of variables and parameters for graphical results [19].

Parameter	Description	Assumed Values
t	Time	0.5
M	Magnetic variable	2
K_1	Dust particles parameter	5
K_2	Dusty fluid variable	1
Gr	Grashof number	2
Pe	Peclet number	1
N	Radiation variable	2
ε	———	0.001
ω	———	$\frac{\pi}{2}$

The obtained results are shown in Figures 2 and 3, respectively, which reflect the Grashof number's behavior on the velocity profile of fluid and dusty particles. We observed the direct variation between the base fluid's velocities and dust particles' velocity in these graphs. By increasing the Grashof number, the rate of fluid and dust particles also increases. According to Grashof number physics, we know that it is the ratio of buoyancy forces and drag forces. Therefore, by increasing this number, the buoyancy forces increase, and the viscosity decreases, which is why the velocities of fluid and dust particles increase. Figures 4 and 5 displayed the fluid and dust particles velocities against the dusty parameter K_2 these graphs show that the fluid and dusty particles' velocity also increases by increasing the dusty parameter. To illustrate the effect of radiation on both the velocities of dusty particles and base fluid, for this behavior, one can observe Figures 6 and 7, respectively. These figures show that when the radiation increases, the velocity of fluid and dust particles also increases. It is evident from the physics of radiation that by increasing the radiation, the fluid temperature and dust particles' kinetic energy are also increased because of this increase in temperature. The corresponding Figures 8 and 9 are plotted respectively for the investigation of the influence of magnetic parameter M on the profile of fluid velocity and also on the velocity of dusty particles. It is noted from these figures that the velocity of the fluid shortens monotonically due to the rise in magnetic parameter M. This reduction in fluid velocity is actually the implementation of magnetic force against the direction of fluid flow. These figures also clear that velocity profiles for the velocity of fluid are much larger than those for dusty particles' velocity. The affiliation of the radiation and the temperature also check out in this article. Figure 10 tells about the relation of radiation variable parameter N and temperature. It is clear from this specific graph that an increase in radiation variable parameter N occurs to a rise in the fluid temperature. Figures 11 and 12 highlight the velocity of the fluid and dust particle in 3D. It shows the better result for the t. Figure 13 is plotted for the comparison of our problem with Narahari and Pendyala [38] which shows a strong agreement. Both solutions are exactly overlapped which shows the correctness and validity of our solutions.

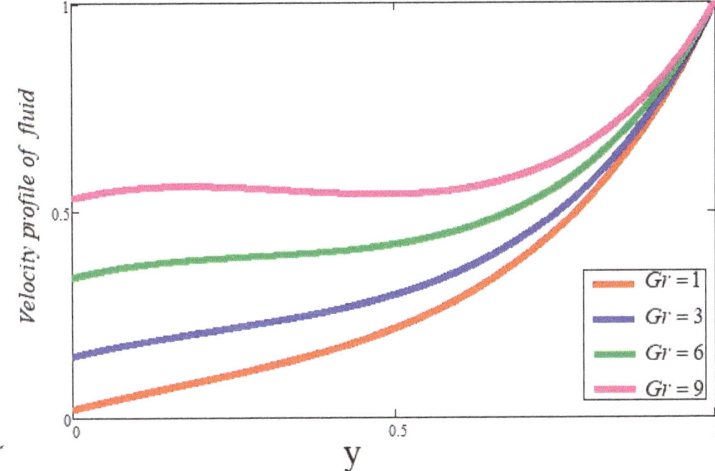

Figure 2. Impact of Gr on the profile of fluid velocity.

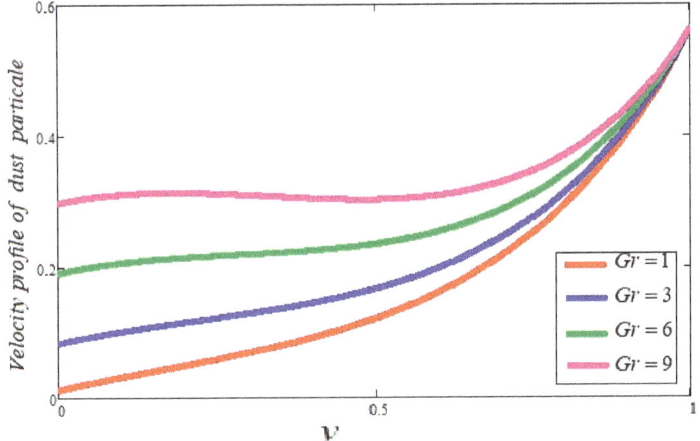

Figure 3. Impact of Gr on the profile of dust particles velocity.

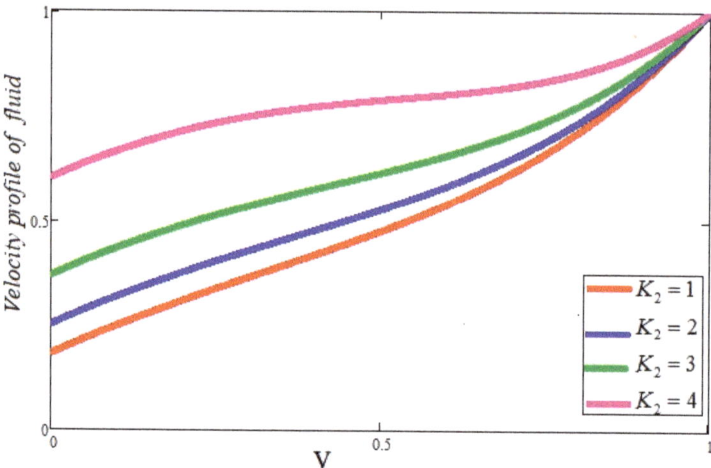

Figure 4. Impact of K_2 on the profile of fluid velocity.

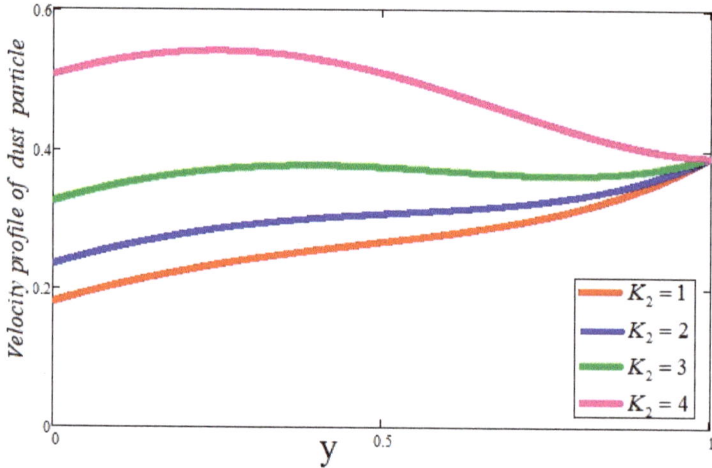

Figure 5. Impact of K_2 on the profile of dust particles velocity.

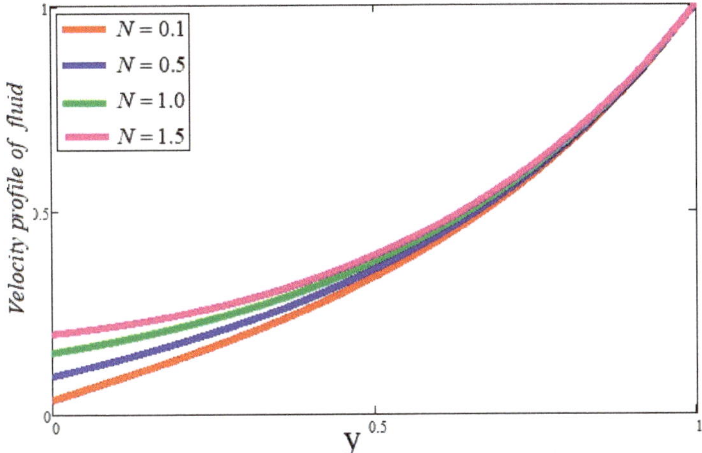

Figure 6. The behavior of N on velocity profile (fluid).

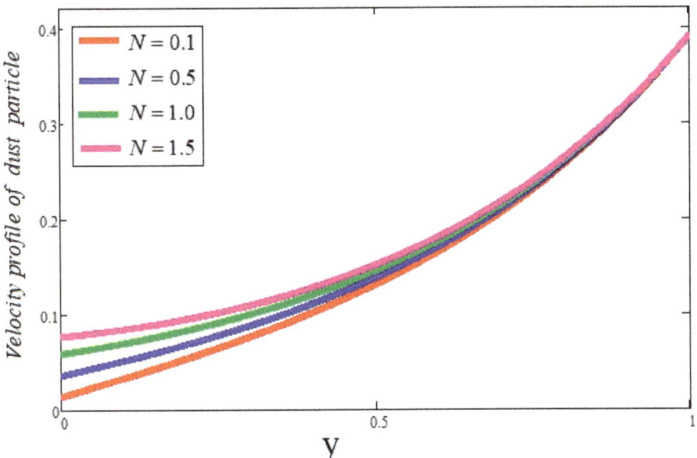

Figure 7. Impact of N on the profile of dust particle velocity.

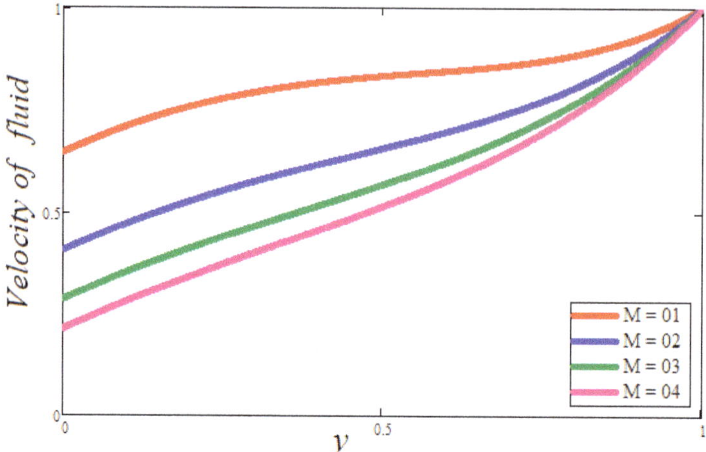

Figure 8. Impact of M on the profile of fluid velocity.

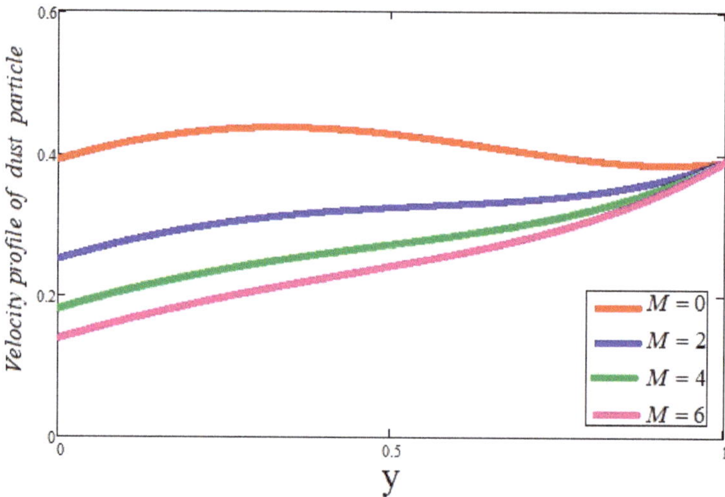

Figure 9. The impact of M on the profile velocity of dust particles.

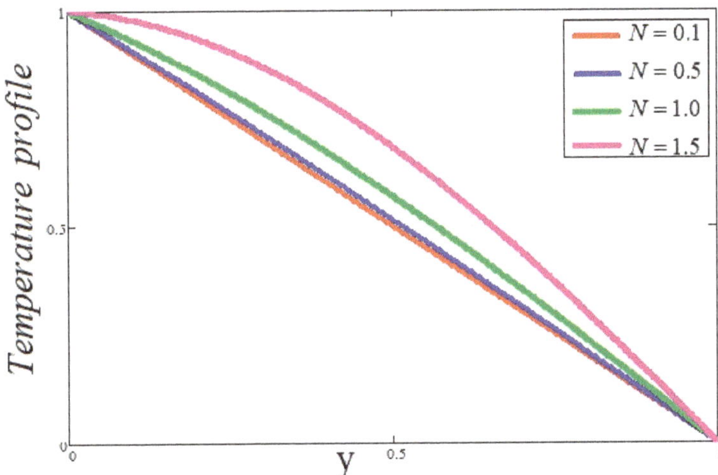

Figure 10. Impact of N on temperature.

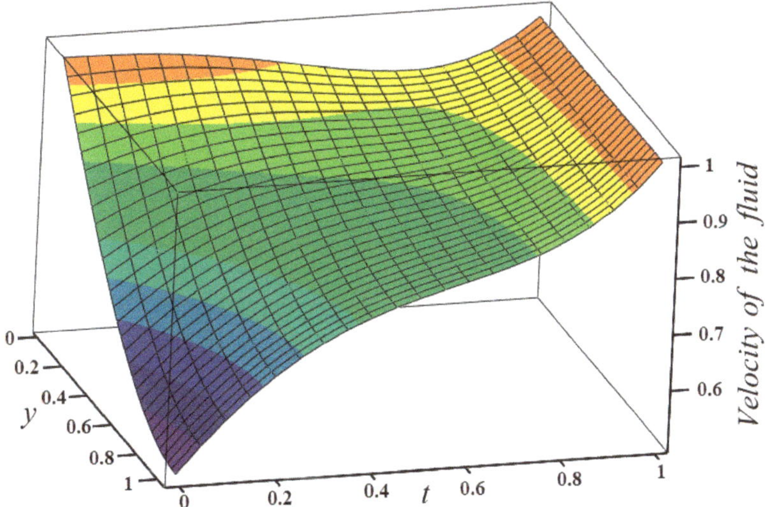

Figure 11. 3D plot of fluid velocity.

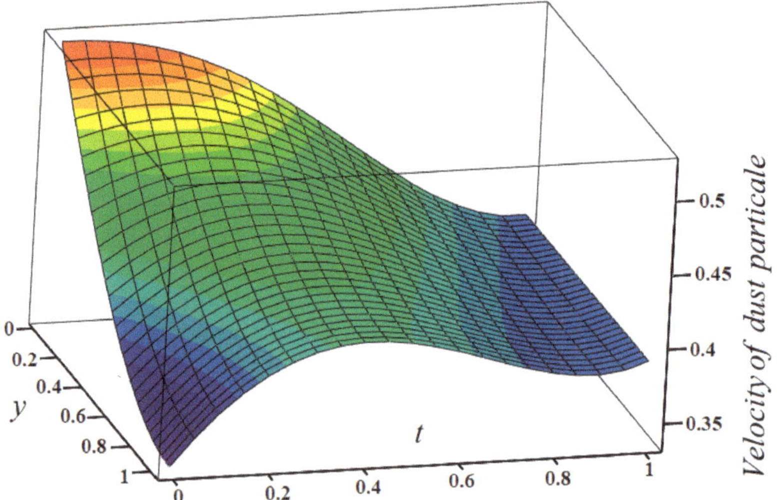

Figure 12. 3D plot of dust particle velocity.

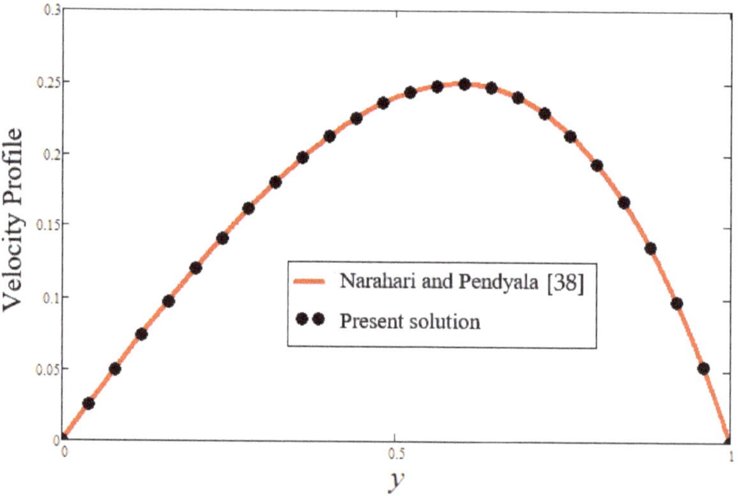

Figure 13. Comparison of the present solutions obtained in Equation (26) when $M = 0, K_1 = 0$, $u(1,t) = 0$ and $K_2 = 0$ with Narahari and Pendyala Equation (11).

5. Conclusions

A novel and theoretical analysis of wall share stress on the consequence of the various physical parameters on the MHD flow of two phases fluctuating flow of dusty fluid is examined along x-direction between two parallel plates, by Poincare-Light Hill Technique. It is pretended that the fluid flow is unidirectional, one dimensional, incompressible, conducting electrically, and heat convection with heat transfer is also appropriated in this problem. The ingrained dust particles are even pretended to be conducting and homogeneously dispersed in the fluid. The parametric consequence of the physical parameters on the profile of fluid velocity, dusty particles and temperature are examined comprehensively. It is observed that increase in Grashof number Gr, dusty parameter K_2, radiation variable parameter N occurs an increase in both the velocities of fluid and dusty particles. The increase in magnetic parameter M occurs a decrease in the velocities of fluid and dusty particles. The relation of radiation and temperature are also discussed graphically in this article and according to this relation the increase in radiation cause the increase in temperature. All the parameters show the identical behaver on fluid and dust particle velocity. While the wall share stress give the verity of velocity profile on the lower plate.

Author Contributions: D.K. model the problem. D.K and G.A. solved the modeled problem analytically. D.K. and G.A. draw the graphs. Results and discussions have reviewed by A.K. and P.K. reviewed the whole manuscript. Methodology, writing—original draft by A.u.R. Proof reading has performed by A.K, I.K. and P.K. All authors have read and agreed to the published version of the manuscript.

Funding: This research was funded by Thailand Science Research and Innovation (TSRI) Basic Research Fund: Fiscal year 2021 under project number 64A306000005 and Chiang Mai University" and "The APC was funded by Research Center of Excellence in Theoretical and Computational Science (TaCS-CoE), KMUTT.

Institutional Review Board Statement: Not applicable.

Informed Consent Statement: Not applicable.

Data Availability Statement: Not applicable.

Acknowledgments: The authors wish to thank the anonymous referees for their comments and suggestions. The authors acknowledge the financial support provided by the Center of Excellence in Theoretical and Computational Science (TaCS-CoE), KMUTT. This research project is supported by Thailand Science Research and Innovation (TSRI) Basic Research Fund: Fiscal year 2021 under project number 64A306000005. Moreover, this research was supported by Chiang Mai University.

Conflicts of Interest: The authors declare no conflict of interest.

References

1. Crowe, C.T.; Troutt, T.R.; Chung, J.N. Numerical models for two-phase turbulent flows. *Ann. Rev. Fluid Mech.* **1996**, *28*, 11–43. [CrossRef]
2. Amkadni, M.; Azzouzi, A.; Hammouch, Z. On the exact solutions of laminar MHD flow over a stretching flat plate. *Commun. Nonlinear Sci. Numer. Simul.* **2008**, *13*, 359–368. [CrossRef]
3. Michael, D.H.; Miller, D.A. Plane parallel flow of a dusty gas. *Mathematika* **1966**, *13*, 97–109. [CrossRef]
4. Healy, J.V. Perturbed Two-Phase Cylindrical Type Flows. *Phys. Fluids* **1970**, *13*, 551–557. [CrossRef]
5. Blevins, R.D. *Applied Fluid Dynamics Handbook*; Krieger Pub: New York, NY, USA, 1984.
6. Ghosh, S.; Hunt, J.C.R. Induced air velocity within droplet driven sprays. *Proc. R. Soc. Lon. Ser. A Math. Phys. Sci.* **1994**, *444*, 105–127.
7. Ahmed, N.; Sarmah, H.K.; Kalita, D. Thermal diffusion effect on a three-dimensional MHD free convection with mass transfer flow from a porous vertical plate. *Lat. Am. Appl. Res.* **2011**, *41*, 165–176.
8. Takhar, H.S.; Roy, S.; Nath, G. Unsteady free convection flow over an infinite vertical porous plate due to the combined effects of thermal and mass diffusion, magnetic field and Hall currents. *Heat Mass Transf.* **2003**, *39*, 825–834. [CrossRef]
9. Wilks, G. Magnetohydrodynamic free convection about a semi-infinite vertical plate in a strong cross field. *Z. Angew. Math. Phys.* **1976**, *27*, 621–631. [CrossRef]
10. Gupta, A.S. Steady and transient free convection of an electrically conducting fluid from a vertical plate in the presence of a magnetic field. *Appl. Sci. Res.* **1960**, *9*, 319. [CrossRef]
11. Gupta, A.S. Laminar free convection flow of an electrically conducting fluid from a vertical plate with uniform surface heat flux and variable wall temperature in the presence of a magnetic field. *J. Appl. Math. Phys.* **1962**, *13*, 324–333. [CrossRef]
12. Soo, S.L. *Fluid Dynamics of Multiphase Systems*; Blaisdell Publishing Co.: Waltham, MA, USA, 1967; Volume 1.
13. Grew, K.N.; Chiu, W.K. A dusty fluid model for predicting hydroxyl anion conductivity in alkaline anion exchange membranes. *J. Electrochem. Soc.* **2010**, *157*, B327. [CrossRef]
14. Saffman, P.G. On the stability of laminar flow of a dusty gas. *J. Fluid Mech.* **1962**, *13*, 120–128. [CrossRef]
15. Vimala, C.S. Flow of a dusty gas between two oscillating plates. *Def. Sci. J.* **1972**, *22*, 231–236.
16. Venkateshappa, V.; Rudraswamy, B.; Gireesha, B.J.; Gopinath, K. Viscous dusty fluid flow with constant velocity magnitude. *Electron. J. Theor. Phys.* **2008**, *5*, 237–252.
17. Zhou, T.; Ge, J.; Shi, L.; Fan, J.; Liu, Z.; Joo, S.W. Dielectrophoretic choking phenomenon of a deformable particle in a converging-diverging microchannel. *Electrophoresis* **2018**, *39*, 590–596. [CrossRef] [PubMed]

18. Zhou, T.; Ji, X.; Shi, L.; Zhang, X.; Song, Y.; Joo, S.W. AC dielectrophoretic deformable particle-particle interactions and their relative motions. *Electrophoresis* **2020**, *41*, 952–958. [CrossRef]
19. Ali, F.; Bilal, M.; Gohar, M.; Khan, I.; Sheikh, N.A.; Nisar, K.S. A Report On Fluctuating Free Convection Flow Of Heat Absorbing Viscoelastic Dusty Fluid Past In A Horizontal Channel With MHD Effect. *Sci. Rep.* **2020**, *10*, 1–15.
20. Ali, F.; Bilal, M.; Sheikh, N.A.; Nisar, K.S.; Khan, I. Two-Phase Fluctuating Flow of Dusty Viscoelastic Fluid between Non-Conducting Rigid Plates with Heat Transfer. *IEEE Access* **2019**, *7*, 123299–123306. [CrossRef]
21. Attia, H.A.; Al-Kaisy, A.M.A.; Ewis, K.M. MHD Couette flow and heat transfer of a dusty fluid with exponential decaying pressure gradient. *J. Appl. Sci. Eng.* **2011**, *14*, 91–96.
22. Ali, F.; Khan, I.; Shafie, S. Closed Form Solutions for Unsteady Free Convection Flow of a Second Grade Fluid over an Oscillating Vertical Plate. *PLoS ONE* **2014**, *9*, e85099.
23. Massoudi, M. Constitutive relations for the interaction force in multicomponent particulate flows. *Int. J. Non-Linear Mech.* **2003**, *38*, 313–336. [CrossRef]
24. Fedkiw, R.P. Coupling an Eulerian fluid calculation to a Lagrangian solid calculation with the ghost fluid method. *J. Comput. Phys.* **2002**, *175*, 200–224. [CrossRef]
25. Anthony, E. Fluidized bed combustion of alternative solid fuels; status, successes and problems of the technology. *Prog. Energy Combust. Sci.* **1995**, *21*, 239–268. [CrossRef]
26. Gidaspow, D. *Multiphase Flow and Fluidization*; Academic Press: New York, NY, USA, 1994.
27. Ungarish, M.; Ungarish, P.D.M. Hydrodynamics of Suspensions. *Hydrodyn. Suspens.* **1993**, *290*, 406.
28. Lighthill, M. CIX. A Technique for rendering approximate solutions to physical problems uniformly valid. *Lond. Edinb. Dublin Philos. Mag. J. Sci.* **1949**, *40*, 1179–1201. [CrossRef]
29. Comstock, C. The Poincaré–Lighthill Perturbation Technique and Its Generalizations. *SIAM Rev.* **1972**, *14*, 433–446. [CrossRef]
30. Große, S.; Schröder, W. High Reynolds number turbulent wind tunnel boundary layer wall-shear stress sensor. *J. Turbul.* **2009**, *1*, N14. [CrossRef]
31. Amili, O.; Soria, J. Wall shear stress distribution in a turbulent channel flow. In Proceedings of the 15th International Symposium on Applications of Laser Techniques to Fluid Mechanics, Lisbon, Portugal, 5–8 July 2010.
32. Örlü, R.; Schlatter, P. On the fluctuating wall-shear stress in zero pressure-gradient turbulent boundary layer flows. *Phys. Fluids* **2011**, *23*, 021704. [CrossRef]
33. Moß, C.; Jarmatz, N.; Hartig, D.; Schnöing, L.; Scholl, S.; Schröder, U. Studying the Impact of Wall Shear Stress on the Development and Performance of Electrochemically Active Biofilms. *ChemPlusChem* **2020**, *85*, 2298–2307. [CrossRef] [PubMed]
34. Hsiao, K.L. To promote radiation electrical MHD activation energy thermal extrusion manufacturing system efficiency by using Carreau-Nanofluid with parameters control method. *Energy* **2017**, *130*, 486–499. [CrossRef]
35. Hsiao, K.-L. Combined electrical MHD heat transfer thermal extrusion system using Maxwell fluid with radiative and viscous dissipation effects. *Appl. Therm. Eng.* **2017**, *112*, 1281–1288. [CrossRef]
36. Hsiao, K.-L. Micropolar nanofluid flow with MHD and viscous dissipation effects towards a stretching sheet with multimedia feature. *Int. J. Heat Mass Transf.* **2017**, *112*, 983–990. [CrossRef]
37. Hsiao, K.-L. Stagnation electrical MHD nanofluid mixed convection with slip boundary on a stretching sheet. *Appl. Therm. Eng.* **2016**, *98*, 850–861. [CrossRef]
38. Narahari, M.; Pendyala, R. Exact Solution of the Unsteady Natural Convective Radiating Gas Flow in a Vertical Channel. In *AIP Conference Proceedings*; American Institute of Physics: University Park, MD, USA, 2013; Volume 1557, pp. 121–124.

Article

Physicochemical Effects of Humid Air Treated with Infrared Radiation on Aqueous Solutions

Olga Yablonskaya [1,*], Vladimir Voeikov [2], Ekaterina Buravleva [2], Aleksei Trofimov [1] and Kirill Novikov [2]

1 N.M. Emanuel Institute of Biochemical Physics, RAS, 119334 Moscow, Russia; avt_2003@mail.ru
2 Faculty of Biology, M.V. Lomonosov Moscow State University, 119234 Moscow, Russia; v109028v1@yandex.ru (V.V.); b_u_k_a@mail.ru (E.B.); kirniknov@yandex.ru (K.N.)
* Correspondence: olga.yablonsky@gmail.com

Abstract: Water vapor absorbs well in the infrared (IR) region of the spectra. On the other hand, it was recently demonstrated that IR radiation promotes formation of the so-called exclusion zones (EZ) at the interfaces between hydrophilic surfaces and water. EZ-water properties differ significantly from that of bulk water. It was studied for the first time whether treatment of water with humid air irradiated with IR-C band could change its physical-chemical properties, making it EZ-water-like. Humid air irradiated with IR was called coherent humidity (CoHu). Redox potential and surface tension decreased in deionized water and mineral water samples that were treated with CoHu, while dielectric constant increased in such water samples. After such treatment of carbonate or phosphate buffers, their buffer capacity against acidification and leaching significantly increased. No such changes were observed in water samples treated with non-irradiated humid air. Thus, after treatment of tested aqueous systems with humid air exposed to IR radiation, their properties change, making them more like EZ-water. The results suggest that IR irradiation of humid air converts it into a carrier of a certain physical signal that affects water properties.

Keywords: water vapor; coherent phase; exclusion zone; redox potential; pH; infrared; microdroplets

Citation: Yablonskaya, O.; Voeikov, V.; Buravleva, E.; Trofimov, A.; Novikov, K. Physicochemical Effects of Humid Air Treated with Infrared Radiation on Aqueous Solutions. *Water* **2021**, *13*, 1370. https://doi.org/10.3390/w13101370

Academic Editors: Maksim Pakhomov and Pavel Lobanov

Received: 26 March 2021
Accepted: 11 May 2021
Published: 14 May 2021

1. Introduction

There is more and more evidence that water, the most common substance in nature and in particular in all living organisms, performs not only the function of a solvent, but also carries certain active functions due to its ability to cooperate and form supramolecular structures with physicochemical properties that differ from those expected for the ideal homogeneous water [1,2]. Indeed, atomic force microscopy allowed to image up to millimeter large stable water clusters consisting of millions of water molecules present in bulk water [3]. An idea of heterogeneous water was also supported by the dynamic light scattering that allowed to detect stable supramolecular complexes in the range of hundreds of nanometers in water that could be formed and reorganized, which at least complements the flickering water clusters model that was suggested earlier [4]. Mae Wan Ho et al. demonstrated that water inside living organisms, unlike bulk water, exhibits properties of a liquid crystal [5]. X-ray emission spectroscopy and neutron diffraction study provided evidence that there is high-density water and low-density water [6,7]. According to these authors, high-density water is characterized by strong hydrogen bonds between molecules surrounded by low-density water with weak hydrogen bonds.

Formation of heterogeneous water is often triggered by particles in water and by surfaces that it wets. The phenomenon of the formation of negatively charged exclusion zones (EZ) in water at hydrophilic surfaces that exclude most solute molecules and have peculiar physicochemical properties has been described years ago [8]. The size of EZ-water varies from tens of nanometers to hundreds of microns, depending on the properties of the surface, quality of water and electromagnetic influences on water [9]. Thus, illumination with infrared (IR) light leads to the increase of EZ as it weakens bonding between water molecules in bulk water adjacent to EZ-water, and "free" water dipoles are attracted to the negatively charged EZ-water region and support its buildup [10]. In another study, the formation of structured water with the use of infrared radiation emitting ceramic powder without its contact with water was demonstrated [11]. As all liquids of a living organism are constantly in contact with hydrophilic surfaces (biopolymeric molecules, supramolecular complexes, cell surfaces), EZ research is important for understanding the fundamental mechanisms of biological action.

Absorption characteristics of IR light by water are determined by rotation and vibration of its molecules. According to Hamashima et al., the spectral signature of a fully hydrated, four-coordinated water molecule is in the hydrogen-bonded O-H stretch region of the IR spectra at 3000–3600 cm^{-1}, depending on the cluster size [12]. In liquid water, hydrogen bonding limits rotational and vibrational changes of water molecules that occur after adsorption of visible and infrared light. In vapor, the gaseous state, rotation of the water molecule is free, combined with stretching and bending of the O-H bonds of water with absorption in the IR spectra region, and leads to a huge number of rotational and vibrational combinations [13,14]. Interestingly, large, hydrogen-bonded water clusters of water molecules make the major contribution to water vapor and humid air absorption in IR regions where individual water molecules do not absorb [15]. The ratio of water liquid-to-vapor IR absorption coefficients can be as high as 10^4. Researchers suggested a hypothesis that water vapor anomalous adsorption can be explained by the presence of areas in vapor rich with hydrogen bonding, which resembles liquid water, whose hydrogen bonds adsorb intensely in the IR region. It is the presence of hydrogen bonds that is responsible for the absorption of electromagnetic energy by clusters in the infrared range. This peculiar adsorption was regarded as an indicator of the presence of water clusters. However, these clusters were considered neutral and uncharged.

There is experimental proof that aqueous systems are capable of storage, modification and transmission of external electro-magnetic signals from a source molecule to biological targets, specifically affecting their endogenous activity and closely resembling the effect of a source molecule via a resonance effect [16]. Here, we irradiate humid air with IR energy with the wavelength of 4000 nm and inquire if such air can change the properties of distilled and mineral bottled waters.

2. Materials and Methods

2.1. Water Preparation

Type I pure deionized water was prepared with Millipore Direct Q3, Merck, and stored in a 500 mL sterile dark borosilicate glass bottle. Senezhskaya and BioVita mineral waters were purchased in 1.5 L plastic bottles. The chemical characteristics of Senezhskaya water were as follows: pH 7.52, [HCO_3^-] 350 mg/L, [Na^+] 5.5 mg/L, [K^+] 11.8 mg/L, [Ca^{2+}] 67 mg/L, [Na^+] 6.9 mg/L, [Mg^{2+}] 27 mg/L, [Cl^-] 39 mg/L, total hardness 5.2 grains/gal, total mineralization 496 mg/L. The chemical characteristics of BioVita water were as follows: pH 7.6, [HCO_3^-] 487 mg/L, [Na^+] 7.4 mg/L, [K^+] 17.3 mg/L, [Ca^{2+}] 94 mg/L, [Na^+] 6.9 mg/L, [Mg^{2+}] 20 mg/L, [Cl^-] 49 mg/L, total hardness 5.5 grains/gal, total mineralization 513 mg/L, and measurements were performed according to standard specifications presented by the American public health association [17]. Water from bottles was transferred in 500 mL sterile dark borosilicate glass bottles. Then, the bottles were left in a dark place at room temperature and were treated with humid air irradiated with IR afterwards.

2.2. Water Treatment

CoHu was produced with a NanoVi Exo (Eng 3 Corporation, Seattle, WA, USA) device that generates humid air with particles 1–10 μm in size out of pure distilled water (Figure 1). The device consists of a humidifier, excitation unit and a control unit. The method involves humidification of the carrier air with water microdroplets, irradiating it with infrared electromagnetic energy and treatment of bulk water or other objects with the irradiated humid air. Unpurified ambient air with initial humidity of 25–30% is used as the carrier gas. The pump (in) takes in ambient air and presses it through the diffuser to water in the humidifier glass container. The pump (out) takes the humidified air above water in the humidifier glass container and presses it through the quartz glass tube of the excitation unit. Each pump creates an air output of 3–4 L per minute. The pump membrane is made of medical grade silicon rubber. The humidified airstream (70–85% humidity) goes to the excitation unit equipped with 24 LEDs, a heat sink element and a thermo sensor. The humid air goes through a 40 mm long quartz glass tube with an inner diameter of 3 mm. Within the glass tube, the water vapor absorbs energy that is emitted by the LEDs placed around the tube at a distance of 2.3 mm. The diameter of each LED is 4.7 mm, and they emit IR energy at about 4000 nm with a beam cone of 30°and the output power of 700 pW. The control unit allows to switch the LEDs off and to obtain the non-irradiated humid air for control measurements. After exiting the excitation unit, humid air exits the device through the 70 cm long flexible plastic outlet with the inner diameter of 3.2 mm.

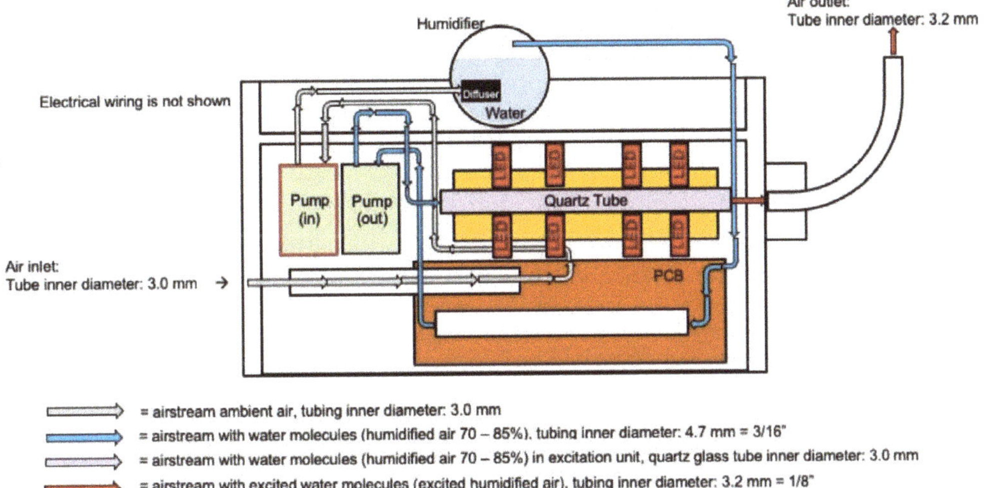

Figure 1. The scheme of the apparatus for CoHu preparation.

Pure and mineral waters were placed in sterile 20 mL glass flasks, and a humid air supply tube was placed 5 cm above the water surface in ambient light. Three parallel 20 mL samples were treated with CoHu and non-CoHu for 1 to 3 min depending on experiment design. Two ways of application of CoHu and non-CoHu were suggested — "blowing" the airstream above the water surface and "bubbling" the airstream via immersion of sterile shortened medical PVC cannula (Philips, Amsterdam, Netherlands) into water samples. Untreated samples were left as controls. Then, the samples were covered with Parafilm, and water parameters were measured within 5 min.

2.3. Redox Potential Measurement

Redox potentials were measured using the potentiometer Ekspert-001 (Ekoniks-Ekspert, Moscow, Russia) equipped with combined Pt/AgCl electrodes in control samples, samples treated with non-CoHu and CoHu for 2 min. In all three groups of samples, redox potential was measured for 10 min continuously and once in an hour during 2 h after treatment, with humid airstream at 21 °C. Measurements were repeated with intervals of 5–10 min.

2.4. Surface Tension Measurement

Surface tension of water samples was measured by using the KSV Instruments Sigma 702 ET tensiometer using the ring method. Surface tension was calculated based on the maximum tension at the moment when a platinum-iridium ring, being slowly pulled out of the liquid sample, broke the surface. Surface tension measurements were repeated 3 times for each group of samples. CoHu or non-CoHu were applied to 20 mL water samples during 5 min.

2.5. Dielectric Constant Measurement

Dielectric constant in pure water and mineral waters, along with samples treated with CoHu and non-CoHu, was measured using the Brookhaven Instruments BI 870 dielectric constant meter at 25 °C with 2% accuracy at 10 kHz. The measurements were repeated 3 times for each group of samples and controls. Treatment with CoHu or non-CoHu was in 10 mL liquid samples during 2 min.

2.6. pH Measurements

Ultra-pure water was obtained as described above. Sodium bicarbonate, HCl, NaOH, Na_2HPO_4 and NaH_2PO_4 were purchased from Sigma-Aldrich (St. Louis, MO, USA).

This experiment had a goal to access the ability of CoHu, the humid air irradiated with IR electromagnetic frequency, to affect buffer capacity of water and buffer solutions. We used pure water, 0.01 M bicarbonate buffer with pH 8.2 and 0.05 M phosphate buffer with pH 7.0. Ultra-pure water was obtained as described above. Sodium bicarbonate, HCl, NaOH, Na_2HPO_4 and NaH_2PO_4 were purchased from Sigma-Aldrich (USA).

Ten mL of a buffer was put into a glass beaker with a magnet, and the beaker was placed on a magnetic mixer. Experimental and control samples were treated with CoHu or with non-CoHu for either 1 or 3 min. Samples were treated with air-humidity using the blowing method without immersing a cannula into the liquid.

Bicarbonate buffer was prepared by dissolving 0.84 g of $NaHCO_3$ in 1000 mL of distilled water. Tests were carried out with each group of samples separately. A 150 μL portion of 0.1 M HCl was added to 0.01 M bicarbonate buffer solution and 30–40 s later, pH value was registered with a pH-meter Ekspert-001-3pH (Ekoniks-Ekspert, Moscow, Russia) (equipped with combined platinum/AgCl electrodes). Then, the next portion of HCl was added. This procedure was continued until 10–11 portions of HCl were added. In another set of tests, 0.1 M NaOH was used for titration. This type of experiment was repeated with 0.05 M phosphate buffer with pH 7.0, and portions of HCl and NaOH were per 750 μL. A part of the samples that was treated with CoHu and non-CoHu was not used in titration immediately, but in 24, 48 and 72 h after treatment.

2.7. Data Processing

Data were collected and analyzed with a PC equipped with Windows 7 Professional software, Microsoft Office Excel 2010 (Microsoft Corp., Redmond, WA, USA), Statistica 10 (Statsoft, Tulsa, OK, USA) and Minitab 17 (Minitab Ltd., Coventry, UK).

3. Results

3.1. Redox Potentials Measurements in Pure and Mineral Waters

Figure 2 illustrates the effect of "blowing" of humid airstream upon water surface of pure water, Senezhskaya and BioVita on their redox potentials. Bubbling of humid air led to either irreproducible or no significant effect, presumably because of excessive perturbation of the aqueous medium. For the same reason, the treatment time was reduced to 2 min in the course of preliminary tests. Redox potential was measured for 10 min continuously, and then the electrode was removed and immersed again into the sample 1 and 2 h after treatment with humid airstream. Control samples were not treated. Comparison of initial redox potential data in different waters shows that pure water has lower redox potential compared to Senezhskaya and BioVita. Samples that were treated with CoHu for 2 min had lower redox potential in all types of water, and this trend remained the same during the whole period of measurements. Treatment with non-CoHu increased redox potential in Senezhskaya and BioVita waters in comparison to respective controls, while in pure water, it resulted in a slight decrease. In Senezhskaya water, redox potential did not differ in pure water and in samples treated with non-CoHu 1 and 2 h after treatment, while according to the Mann–Whitney test, there was significant difference between potentials in these waters and water treated with CoHu.

Table 1. Mean surface tension and standard deviation of the mean in pure water, Senezhskaya and BioVita mineral waters in samples that were treated by CoHu, non-CoHu and untreated control.

	Control	CoHu	Non-CoHu
Pure Water Mean, σ	70.29	70.16	70.28
Pure Water σ, SD	0.047	0.044	0.038
Senezhskaya Mean, σ	73.72	73.66	73.81
Senezhskaya σ, SD	0.048	0.042	0.036
BioVita Mean, σ	73.18	73.09	73.18
BioVita σ, SD	0.04	0.047	0.045

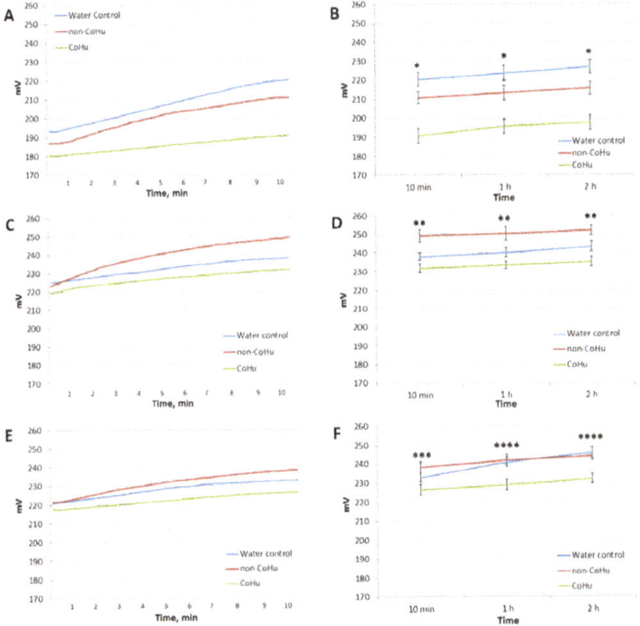

Figure 2. (**A**) A typical continuous measurement of redox potential in pure water in control samples (blue lines), samples Table 1. min after treatment. (**B**) Mean redox potential data recorded 1 and 2 h after treatment are presented. * $p < 0.05$ between the three groups of samples by paired comparison by Mann–Whitney U-test. (**C**) A typical continuous measurement of redox potential in BioVita. (**D**) Mean redox potential data in BioVita recorded 1 and 2 h after treatment. ** $p < 0.05$ samples treated with CoHu differ from the two other groups. (**E**) A typical continuous measurement of redox potential in Senezhskaya mineral bottled water. (**F**) Mean redox potential data in Senezhskaya recorded 1 and 2 h after treatment. *** $p < 0.05$ samples of all groups significantly different from each other, **** $p < 0.05$ samples treated with CoHu differ from the two other groups.

3.2. Surface Tension Measurement

Surface tension was measured in 5 parallel samples for each of the 3 groups in pure water, Senezhskaya and BioVita waters. The results are shown in Table 1. In all water samples, surface tension was unchanged after treatment with non-CoHu for 3 min. Treatment with CoHu led to a slight decrease in surface tension values in all three types of water, especially in BioVita.

According to Fisher pairwise comparisons, mean surface tension in pure water and BioVita controls was significantly different from that of samples treated with CoHu. In Senezhskaya water, the difference was not significant enough. In all water types, there was a reliable difference between samples treated with CoHu and non-CoHu. In neither type of water were differences considered significant between control untreated samples and non-CoHu-treated samples. The data analysis is shown in Figure 3.

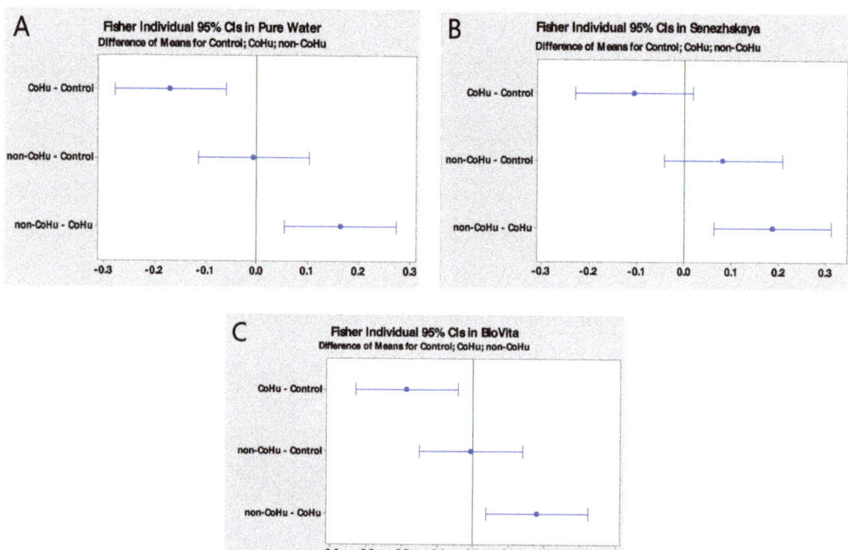

Figure 3. Fisher pairwise comparisons for surface tension measurements in pure water (**A**), Senezhskaya (**B**) and BioVita (**C**) mineral waters, where CIs are confidence intervals. If an interval does not contain zero, the corresponding means are significantly different.

3.3. Dielectric Constant Measurement

The static dielectric constant of pure water is around 81 due to the fact that water is a strongly polar liquid; however, its molecules have a degree of freedom and rotation. Each water molecule has a significant dipole moment. In the absence of an electric field, the dipoles are oriented randomly, and the total electric field created by them is equal to zero. If water is placed in an electric field, then the dipoles will begin to reorient themselves so as to weaken the applied field. Such a picture is observed in other polar liquids, but water, due to the large value of the dipole moment of H_2O molecules, is capable of a very strong (around 80 times) weakening of the external field. Thus, dielectric constant reflects the capability of water to react to external electromagnetic fields. In the presence of solvents, it slightly varies.

Figure 4A presents dielectric constant measurement data and Fisher pairwise comparisons for pure water control samples, samples treated with CoHu and non-CoHu (Figure 4B). Dielectric constant increased in samples that were treated with CoHu and did not significantly differ in samples that were treated with non-CoHu according to the Mann–Whitney U-test. A very similar pattern was observed in Senezhskaya water (Figure 5B) and BioVita (Figure 6B). In both these cases, treatment with CoHu significantly increased the dielectric constant. However, according to Fisher pairwise comparisons, means are different in all pairs in pure water and Senezhskaya (Figures 5B and 6B). In BioVita, mean dielectric constant values did not differ significantly between control and samples treated with non-CoHu (Figure 6B).

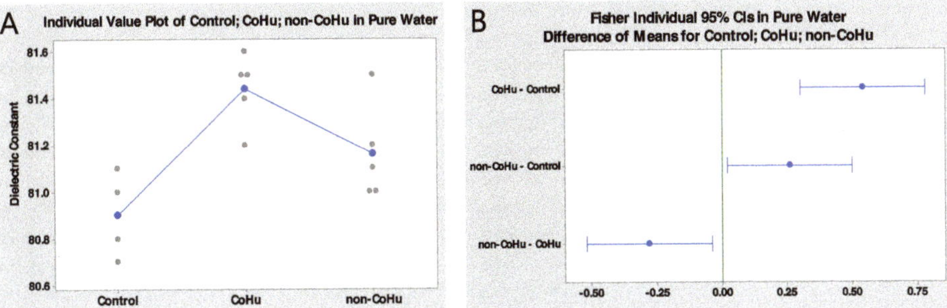

Figure 4. (**A**) Pure water dielectric constant scattered data in samples treated with CoHu and non-CoHu. (**B**) Fisher pairwise comparisons for dielectric constant measurements in pure water. CIs are confidence intervals. If an interval does not contain zero, the corresponding means are significantly different.

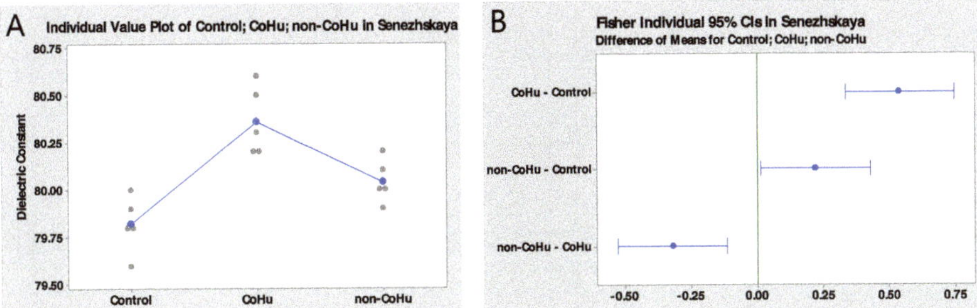

Figure 5. (**A**) Senezhskaya water dielectric constant scattered data in samples treated with CoHu and non-CoHu. (**B**) Fisher pairwise comparisons for dielectric constant measurements in Senezhskaya natural water. CIs are confidence intervals. If an interval does not contain zero, the corresponding means are significantly different.

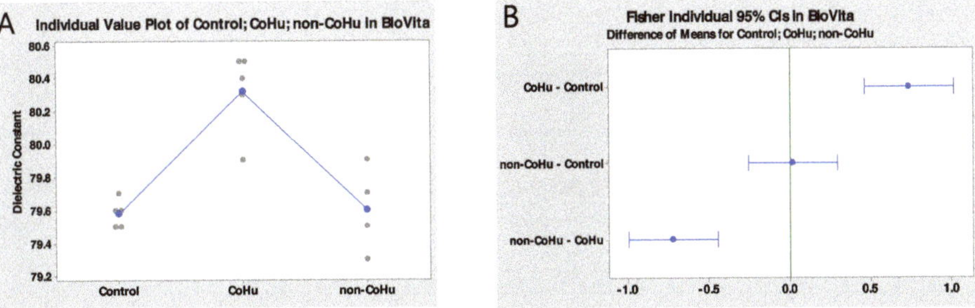

Figure 6. (**A**) BioVita water dielectric constant scattered data in samples treated with CoHu and non-CoHu. (**B**) Fisher pairwise comparisons for dielectric constant measurements in BioVita natural water. CIs are confidence intervals. If an interval does not contain zero, the corresponding means are significantly different.

3.4. pH Measurement

Figure 7 represents typical results of pH measurement in bicarbonate buffer that was titrated with 0.1 mol/L HCl or 0.1 mol/L NaOH. Graphs also show pH values for titration of buffer samples that were treated with CoHu for 3 min 24 h before titration. In case of titration with HCl, pH started to drop after the 7th portion of HCl in control samples and samples that were treated with non-CoHu, while in samples that were treated with CoHu for 1 or 3 min, pH started to drop after the 8th or the 9th portion. In the case of titration with NaOH, pH began to rise in samples that were treated with CoHu after addition of one or two portions, more than it did in control samples and samples treated with non-CoHu. Titrations with HCl and NaOH were repeated 4 times for all groups of samples. In the case of titration with HCl, pH values in samples treated with CoHu after the addition of the 8th and 9th portion significantly differed from control samples and samples treated with non-CoHu according to the Mann–Whitney test, with $p = 0.0128$ and 0.0117, respectively. Similarly, in the case of titration with NaOH, pH values in samples treated with CoHu significantly differed on the addition of the 7th and 8th portion of alkali from control samples and samples treated with non-CoHu, with $p = 0.0376$ and 0.0202, respectively. After addition of the 9th portion of NaOH, pH data was reliably different between samples that were treated with CoHu for 3 min before titration and 24 h before titration ($p < 0.05$). pH values in control samples and samples treated with non-CoHu did not differ.

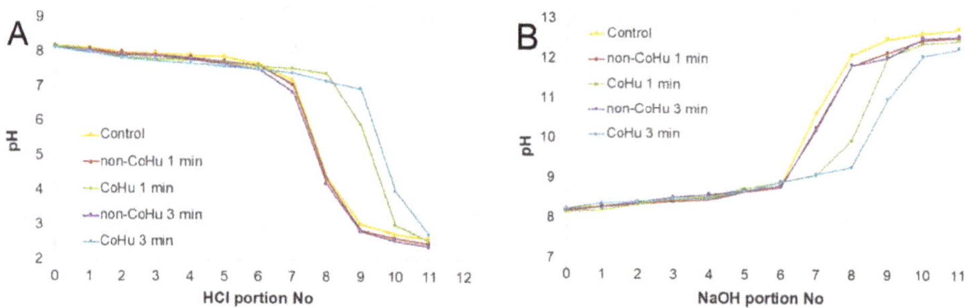

Figure 7. (**A**) A typical curve of titration of 0.01 M bicarbonate buffer treated with CoHu or non-CoHu with 0.1 M HCl and (**B**) with 0.1 M NaOH. Some samples were treated with the airstream for 3 min.

In a parallel experiment, pH was measured in the course of titration either with HCl or NaOH of buffer samples that were treated with CoHu or non-CoHu immediately before titration and 1–3 days prior to it. The results with statistical errors are presented in Figures 8 and 9 along with Mann–Whitney U-test probability analysis.

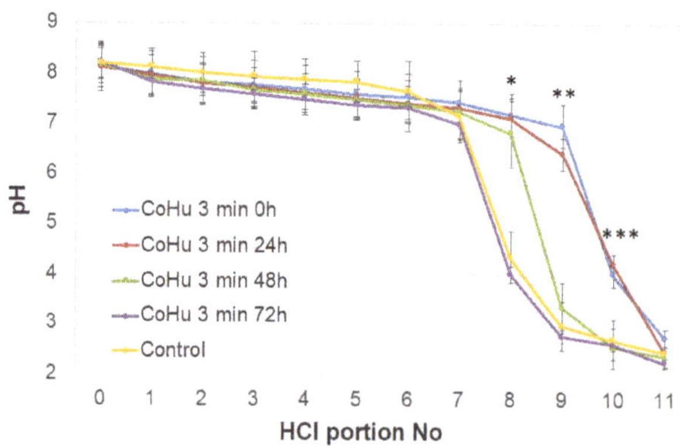

Figure 8. Averaged 0.01 M bicarbonate buffer titration curves with 0.1 M HCl. Samples were treated with CoHu for 3 min immediately before titration, 24, 48 and 72 h before titration. * $p < 0.01$ between control and samples that were treated with CoHu 0, 24 and 48 h before treatment; **, *** $p < 0.01$ between united control and samples that were treated with CoHu 48 h before treatment and samples that were treated with CoHu 0 and 24 h before treatment.

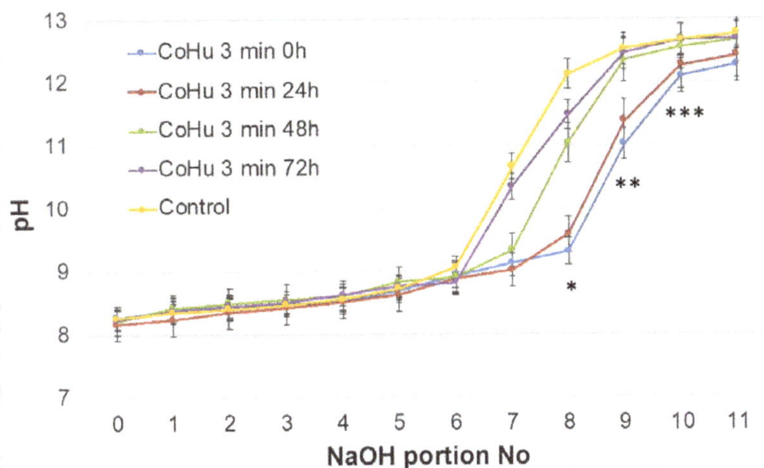

Figure 9. Averaged 0.01 M bicarbonate buffer titration curves with 0.1 M NaOH. Samples were treated with CoHu for 3 min immediately before titration, 24, 48 and 72 h before titration. *, ** $p < 0.01$ between samples that were treated with CoHu immediately and 24 h before treatment and united control samples and samples that were treated with CoHu 48 and 72 h before treatment; *** $p < 0.05$ between control and samples that were treated with CoHu immediately before treatment.

Very similar results were obtained with phosphate buffer (Figure 10). In case of titration with portions of 0.1 M HCl in samples treated with CoHu, pH began to drop 2 or 3 portions later than it did in control samples. Values of pH remained more stable in samples that were treated for 3 min. This effect cannot be attributed to any impact of the airflow as pH change pattern in samples that were treated with non-coherent humid air (non-CoHu) is undistinguishable from controls. Four titrations both with HCl and NaOH were performed.

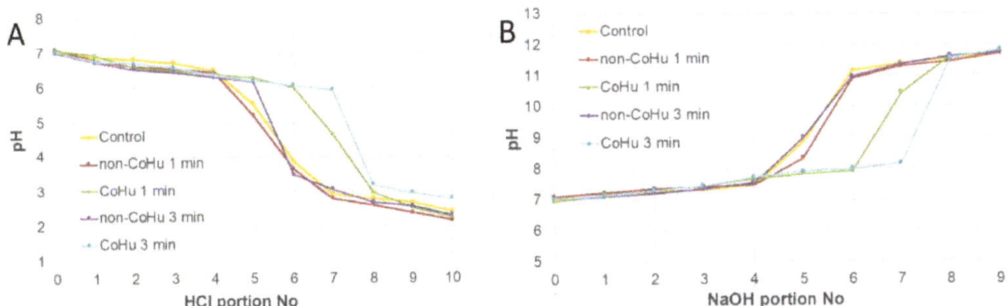

Figure 10. Typical curve of titration of 0.05 M phosphate buffer treated with CoHu or non-CoHu with 0.1 M HCl (**A**) and with 0.1 M NaOH (**B**). Some samples were treated with the airstream for 1 or 3 min.

4. Discussion

According to our knowledge, we examined the effect of coherent humid air on such parameters of the aqueous medium as oxidation-reduction potential, changes in pH with the addition of acid or alkali, surface tension and dielectric constant for the first time. Deionized pure water, two types of mineral waters and buffers were the examples of aqueous systems that were considered as test systems to determine possible effects of CoHu for its potential future application to biological objects, for agricultural and medical purposes. A feature of our experiments is that the water systems were not directly irradiated with infrared light, but aqueous aerosol was used as an intermediary.

Aqueous aerosol, or humid air, produced by a compressor-type evaporator, was treated with IR radiation (CoHu) or was not treated (non-CoHu) for control. This allowed to perform tests with an appropriate control treatment on the impacts of the airstream, shaking and heating, while sample treatments were equal both in test and control samples, which is important for such study that requires accuracy to detect subtle effects.

The volumes of water samples treated with CoHu or non-CoHu were 10 and 20 mL, depending on the method of measurement. In both cases, the surface area to volume ratio was approximately the same, however the study of the effects of volume and surface area is worthy of a separate study.

Redox potential was measured in pure water and the two types of mineral waters—Senezhskaya and BioVita (Figure 2). Untreated control pure water had lower redox potential compared to untreated control mineral waters. Treatment of all types of water with CoHu for 2 min led to a redox potential decrease in all types of water, and this trend remained the same during the whole measurement. In samples that were treated with non-CoHu, redox potential increased in Senezhskaya and BioVita waters, while in pure water, a slight decrease was observed. The redox potential decrease indicates that water treated with CoHu may possess antioxidant properties. The redox potential of the aqueous medium in living organisms is usually lower than in environmental water. Interestingly, redox potential of EZ water formed at the surface of hydrophilic powder also showed a decrease [11]. The development of a negative electric potential of −120 to −160 mV across the boundary between the exclusion zone formed at the Nafion® surface and bulk water (EZ-water is negatively charged) was reported in the context of EZ-water studies [9]. The charge separation occurred immediately as the EZ was formed and protons were excluded from it.

After treatment of pure water (Figure 4), Senezhskaya (Figure 5) and BioVita (Figure 6) waters with CoHu, a tendency for the increase of the dielectric constant in all these waters was observed. Fisher pairwise comparison showed also that in pure water and Senezhskaya water, both CoHu-treated samples and non-CoHu-treated samples had higher mean dielectric constant values. In BioVita water, both CoHu-treated samples and non-CoHu-treated samples differed from untreated water however they did not differ from each other. Dielectric constant indicates the ability of the liquid to weaken the external field. Pure and mineral waters interact with the field and weaken it almost 80 times. The increase of the dielectric constant was slight, though may be indicative of the increased charge storage capacity and the dipole moment of the whole sample or some zones in them.

It should be noted that tap water in which surface tension reaches ~75 dyn/cm hardly supports metabolism in living organisms and in cells. A cell is capable to use water having surface tension of ~43–45 dyn/cm (45 dyn/cm is a biologically optimal surface tension of the tissue fluid and blood) [18]. Surface tension decreased in all types of waters after treatment with CoHu (Table 1), which is in line with [11] as the density of coherent water is approximately 0.97 g/cm^3, as coherent water molecules take up more space. In samples that were treated with non-CoHu, no change in surface tension was observed, so the impact of "blowing" of the airstream upon the sample surface should be considered (Figure 3). More research is needed to determine whether the density is heterogenous or not in the samples treated with CoHu.

The character of pH change in CoHu-treated bicarbonate and phosphate buffer in response to the addition of portions of alkali or acid allows to suggest the ability of CoHu to alter buffer capacity of aqueous systems. Interestingly, this effect did not depend on whether CoHu was applied for 1 or 3 min. Moreover, this effect persisted for 48 h after treatment with CoHu. If it is suggested that CoHu acts as a source of EZ or as a factor providing for its formation in treated water, it can be assumed that the effect on the buffer properties of solutions is associated with the separation of negative and positive charges. We have already demonstrated that in highly diluted hydrated fullerene C60 solutions prepared using vigorous agitation at each step, pH response to the addition of HCl demonstrates an increase in buffering capacity in comparison with control [19]. In the same study, we observed structural heterogeneity in samples with highly diluted hydrated fullerene. This corresponds with data obtained in the laboratory of Elia, who have demonstrated that homeopathic remedies such as Arnica Montana, Arsenicum Album and Magnesium Muriaticum have higher capability to buffer pH decrease when they are titrated with HCl than water used for the preparation of these solutions [20]. The authors suggested that excessive buffering capacity was provided by supramolecular organization occurring in the extremely diluted solutions, namely by the appearance in them of dissipative coherent water structures in the course of serial dilutions with intensive shaking.

Thus, water, either liquid or gaseous, has a potential to form unstable aggregates. Del Giudice and Preparata [21,22] suggested a quantum-physics-based model that involved the presence of 100 nanometers large coherent domains (CDs) in liquid water. Del Giudice and colleagues [23–25] saw similarities between EZ-water and coherent domains, although Pollack [26] has proposed a different structure and origin for EZ-water. The surface charge of the EZ was suggested to depend on the surface charge of the hydrophilic surface on which it is formed, thus positively charged hydrophilic gels promote a positive charge at the surface of the EZ with a high pH zone [27].

Conventional quantum electrodynamic (QED) field theory applies only to gases, however Del Giudice and Preparata expanded it to the condensed phase of liquids. According to QED theory, two substances are formed in water in order to minimize potential energy—CDs and common bulk water. In a CD, all water molecules are in a coherent state with in-phase wave functions. As a result, the overall wave function of the entire domain is a million times magnified wave function of any of the individual water molecules in the domain [28].

The CD contains a large number of quasi free electrons. The reason for this lies in the fact that the energy of the excited state of the CD is 12.06 eV, which is very close to the ionization energy of water molecules of 12.6 eV [29]. Therefore, water in CD can serve as a donor of electrons, which participate in redox processes, and this is in line with the findings of this paper. Emerging and reforming CDs surrounded by bulk incoherent water may play a particular role in the biological functions of water [30–33].

In a study on aerosols formed at waterfalls, Madl et al. have shown [34] the presence of negatively charged nanometer-sized and up to 100 nm large water clusters that contain millions of water molecules. Unlike unstable small clusters, large water aggregates can be found hundreds of meters away from the waterfall. Both negative surface charge and the size are in accordance with the QED theory of water developed by Del Giudice and colleagues [25]. The authors suggest that the river flow offers conditions for formation of CDs made up by both positively and negatively charged entities. Then, it becomes fragmented by the waterfall, acquires a portion of energy and the CDs become separated in the form of an aerosol. Thus, these aerosol water clusters can act as a surface, and exposure to IR is expected to cause the growth of these clusters in humid conditions.

Water clusters can make crucial contributions to the IR absorption spectrum of water vapor [35]. Moreover, infrared absorption of water vapor is related to ion-production in vapor and electrical properties of air, which also fits into the concept of charged surfaces and charge separation during the formation of CDs and EZ-water. Carlson's suggestion that the hydronium ion could be implicated in the absorption at 3–5 μm observed in water vapor measurements also implies charge separation in vapor [36].

Using the methods of dynamic light scattering, nanoparticle tracking analysis (NTA) and transmission electron microscopy (TEM), it was possible to experimentally show the existence of nanoscale self-regulating molecular ensembles in aqueous systems, nano-associates, with sizes of 100–400 nm and a ζ-potential from -2 to -20 mV [37,38]. A necessary condition for their formation is the presence of traces of impurities in the water and the presence of external electromagnetic fields. According to the research data of other authors, from the standpoint of QED, it is possible to assume the formation of nanoscale CDs in the presence of external fields in an aqueous medium [39].

A noteworthy NMR study [40], where proton relaxation times T_1 (the spin-lattice relaxation time) and T_2 (the spin-spin relaxation time) were measured in distilled water, homeopathic remedy, spring water and water treated with electromagnetic fields, demonstrated evidence of the presence of supramolecular structures similar to liquid crystals, whose molecules are arranged in some order and whose state of matter is between that of a liquid and a crystalline solid. The authors suggest that their findings could be explained from the point of the hypothesis of CDs and EZ-water formation.

Messori et al. provided a holistic overview of the role and physical mechanisms of action of EZ and CDs in living organisms, where EZ is regarded as long-range ensembles of CDs [40]. They focused on phase transition of water from the ordinary coherence of its liquid state (bulk water, as water cannot be completely incoherent) to the semi-crystalline state of interfacial water and its role in living organisms, where electron/proton dynamics and response to electromagnetic fields are used to receive electromagnetically encoded signals endowed with coherence at a low frequency. The resultant excitations are summed, and the coherence is distributed at frequencies that may affect biological systems.

It has recently been shown that laser radiation induced ROS (hydrogen peroxide, hydroxyl and superoxide radicals) generation in solutions of blood serum proteins, bovine serum albumin and gamma-globulin, resulting in the formation of long-lived reactive protein species [41]. These visible light- and heat-induced long-lived reactive protein products can generate hydrogen peroxide in aqueous medium and it contributes to the adaptation of living organisms to stress factors [42]. Authors suggest that ROS generation can be related to the release of extra free energy via surface tension in air nanobubbles present in water (bubstons—bubbles stabilized by ions) as a result of their collapse under the action of visible light, laser irradiations or heat. A resonance excitation of molecular oxygen by wavelengths corresponding to its transition to singlet state leads to electromagnetic disturbance, resulting in collapse of nanobubbles. Interestingly, the size of mesoscopic droplets formed in low-concentration aqueous solutions of polar organic compounds increases with temperature [43], which also contributes to the mass of evidence of the peculiar role of heat in water solutions. We have already mentioned the work of [11], where EZ-like water was obtained by its non-contact treatment with IR–emitting QELBY ceramic powder (Quantum Energy Co Ltd., Hanam-si, South Korea) that was at temperature equilibrium with water. Thus, the authors suggest that it was not heating that changed water structure but the IR emission of the powder, which is suggested to be more coherent compared to a more chaotic IR component in environmental IR radiation associated with the ambient heat. The physicochemical characteristics of IR-altered water in this paper are supported by strong evidence of the authors of [44], who have shown that water prepared either by mixture with the ceramic powder, or which is especially noteworthy, by non-contact treatment, exhibited antioxidant properties, stimulated plant growth, increased normal cell culture viability and decreased cancer cells' viability, and also increased the cytokine expression in the splenocytes.

Hydrogen bonding, whose behavior and fluctuations depend on energetic characteristics of O–H bond in water molecules, plays an important role in EZ-water maintenance. Thus, energy absorption by O–H bonds can affect hydrogen bonding and EZ formation.

The question may arise as to why the humid air was treated with IR radiation, and not the liquids themselves. According to Pollack's theory [26], water microdroplets have an exclusion zone water layer on their surface, and water is spherically confined there due to surface tension at the air–water interface. Smaller droplets have a relatively larger EZ area. As it was mentioned above, EZ expands upon absorption of a sufficient amount of IR energy to modify the hydrogen bonding network in a water microdroplet. As water vapor absorbs more intensely in IR, it is used as an inducer of EZ in bulk liquid water. Moreover, irradiation of water with IR leads to its heating, and microdroplets lose obtained heat quicker. Irradiation with LED is mostly coherent and transferring of this coherent signal to humid air with suspended microdroplets and molecular water, when they begin to oscillate in a coherent manner, is more effective than transferring it to liquid water. If we suggest that water microdroplets uptake extra energy in the form of IR and store it in the form of EZ, when they fuse with condensed phase, they are likely to give this energy off. Another aspect comes from studies of plasma–water interactions, where the highest production of hydrogen peroxide in water was obtained by treating water microdroplets with plasma, which is known to cause reactive oxygen species generation in water [45]. This is explained by the large surface area of the vapor droplets. So, it can be assumed that microdroplets in the humid air have an increased surface-to-volume ratio and absorb well in the IR region of the spectrum, which leads to an increase in the EZ and promotes the transition of water molecules to a coherent phase. These microdroplets come into contact with the treated liquid, which leads to a change in its physicochemical parameters. To sum up, water microdroplets are more effective in exchange of IR energy than condensed water.

5. Conclusions

According to the results of our study, in which we investigated the changes in the physicochemical properties of aqueous systems after non-chemical treatment with humid air exposed to IR waves, we obtained a number of similarities with exclusion zone water, which were absent after treatment with untreated humid air and untreated water control. These were changes in the redox potential and dielectric constant. A change in buffer properties indicates an alteration in the distribution of positive and negative charges characteristic of EZ, and a change in surface tension indicates the presence of heterogeneity in the treated aqueous medium, which is typical of both EZ and CDs.

Author Contributions: Conceptualization, O.Y. and V.V.; methodology, O.Y.; software, E.B.; validation, A.T.; investigation, O.Y., K.N. and E.B.; resources, A.T.; writing, O.Y. and V.V.; visualization, E.B. All authors have read and agreed to the published version of the manuscript.

Funding: This research received no external funding.

Institutional Review Board Statement: Not applicable.

Informed Consent Statement: Not applicable.

Data Availability Statement: The data presented in this study are available on request from the corresponding author.

Conflicts of Interest: The authors declare no conflict of interest.

References

1. Cooke, R.; Kuntz, I.D. The properties of water in biological systems. *Annu. Rev. Biophys. Bioeng.* **1974**, *3*, 95–126. [CrossRef]
2. Ling, G.N. The physical state of water in living cell and model systems. *Ann. N. Y. Acad. Sci.* **1965**, *125*, 401–417. [CrossRef]
3. Lo, A.; Cardarella, J.; Turner, J.; Lo, S.Y. A soft matter state of water and the structures it forms. *Forum. Immunopathol. Dis. Ther.* **2012**, *3*, 237–252. [CrossRef]
4. Konovalov, A.I.; Ryzhkina, I.S. Highly diluted aqueous solutions: Formation of nano-sized molecular assemblies (nanoassociates). *Geochem. Int.* **2014**, *52*, 1207–1226. [CrossRef]
5. Ho, M.W.; Haffegee, J.; Newton, R.; Zhou, Y.M.; Bolton, J.S.; Ross, S. Organisms as polyphasic liquid crystals. *Bioelectrochem. Bioenerg.* **1996**, *41*, 81–91. [CrossRef]
6. Tokushima, T.; Harada, Y.; Takahashi, O.; Senba, Y.; Ohashi, H.; Pettersson, L.G.; Nilsson, A.; Shin, S. High resolution X-ray emission spectroscopy of liquid water: The observation of two structural motifs. *Chem. Phys. Lett.* **2008**, *460*, 387–400. [CrossRef]
7. Soper, A.K.; Ricci, M.A. Structures of high-density and low-density water. *Phys. Rev. Lett.* **2000**, *84*, 2881. [CrossRef]
8. Zheng, J.M.; Pollack, G.H. Long-range forces extending from polymer-gel surfaces. *Phys. Rev. E* **2003**, *68*, 031408. [CrossRef] [PubMed]
9. Zheng, J.M.; Chin, W.C.; Khijniak, E.; Khijniak, E., Jr.; Pollack, G.H. Surfaces and interfacial water: Evidence that hydrophilic surfaces have long-range impact. *Adv. Colloid Interface Sci.* **2006**, *127*, 19–27. [CrossRef]
10. Chai, B.; Yoo, H.; Pollack, G.H. Effect of radiant energy on near-surface water. *J. Phys. Chem. B* **2009**, *113*, 13953–13958. [CrossRef] [PubMed]
11. Hwang, S.G.; Hong, J.K.; Sharma, A.; Pollack, G.H.; Bahng, G. Exclusion zone and heterogeneous water structure at ambient temperature. *PLoS ONE* **2018**, *13*, e0195057. [CrossRef] [PubMed]

12. Hamashima, T.; Mizuse, K.; Fujii, A. Spectral signatures of four-coordinated sites in water clusters: Infrared spectroscopy of phenol−(H_2O) n (∼20 ≤ n≤ ∼50). *J. Phys. Chem. A* **2011**, *115*, 620–625. [CrossRef] [PubMed]
13. Lemus, R. Vibrational excitations in H_2O in the framework of a local model. *J. Mol. Spectrosc.* **2004**, *225*, 73–92. [CrossRef]
14. Röttgers, R.; McKee, D.; Utschig, C. Temperature and salinity correction coefficients for light absorption by water in the visible to infrared spectral region. *Opt. Express* **2014**, *22*, 25093–25108. [CrossRef] [PubMed]
15. Carlon, H.R. Aerosol spectrometry in the infrared. *Appl. Opt.* **1980**, *19*, 2210–2218. [CrossRef]
16. Tiezzi, E.; Catalucci, M.; Marchettini, N. The supramolecular structure of water: NMR studies. *Int J. Des. Nat. Ecodyn.* **2010**, *5*, 10–20. [CrossRef]
17. *Standard Methods for the Examination of Water and Wastewater*, 20th ed.; American Public Health Association, American Water Work Association, Water Environment Federation: Washington, DC, USA, 1998.
18. Goncharuk, V.V.; Kovalenko, V.F.; Zlatskii, I.A. Comparative analysis of drinking water quality of different origin based on the results of integrated bioassay. *J. Water Chem. Technol.* **2012**, *34*, 61–64. [CrossRef]
19. Voeikov, V.; Yablonskaya, O.; Buravleva, E.; Novikov, K. Peculiarities of the physicochemical properties of hydrated C60 fullerene solutions in a wide range of dilutions. *Front. Phys.* **2021**, *9*, 64.
20. Elia, V.; Napoli, E.; Niccoli, M. Physical–chemical study of water in contact with a hydrophilic polymer: Nafion. *J. Therm. Anal. Calorim.* **2013**, *112*, 937–944. [CrossRef]
21. Arani, R.; Bono, I.; Del Giudice, E.; Preparata, G. QED coherence and the thermodynamics of water. *Int. J. Mod. Phys. B* **1995**, *9*, 1813–1841. [CrossRef]
22. Del Giudice, E.; Preparata, G. Coherent electrodynamics in water. In *Fundamental Research in Ultra High Dilution and Homoeopathy*; Schulte, J., Endler, P.C., Eds.; Springer: Dordrecht, The Netherlands, 1998; pp. 89–103.
23. Del Giudice, E.; Spinetti, P.R.; Tedeschi, A. Water dynamics at the root of metamorphosis in living organisms. *Water* **2010**, *2*, 566–586. [CrossRef]
24. Del Giudice, E.; Tedeschi, A.; Vitiello, G.; Voeikov, V. Coherent structures in liquid water close to hydrophilic surfaces. *J. Phys. Conf. Ser.* **2013**, *442*, 012028. [CrossRef]
25. Capolupo, A.; Del Giudice, E.; Elia, V.; Germano, R.; Napoli, E.; Niccoli, M.; Tedeschi, A.; Vitiello, G. Self-similarity properties of nafionized and filtered water and deformed coherent states. *Int. J. Mod. Phys. B* **2014**, *28*, 1450007. [CrossRef]
26. Pollack, G.H. *The Fourth Phase of Water*; Ebner & Sons Publishers: Seattle, DC, USA, 2013.
27. Zheng, J.M.; Wexler, A.; Pollack, G.H. Effect of buffers on aqueous solute-exclusion zones around ion-exchange resins. *J. Colloid Interface Sci.* **2009**, *332*, 511–514. [CrossRef] [PubMed]
28. Del Giudice, E.; Preparata, G.; Vitiello, G. Water as a free electric dipole laser. *Phys. Rev. Lett.* **1988**, *61*, 1085. [CrossRef]
29. Voeikov, V.; Korotkov, K. *Novaya Nauka o Vode*; Mednii Vsadnik: St. Petersburg, Russia, 2017; pp. 67–70. (In Russian)
30. Del Giudice, E. Old and new views on the structure of matter and the special case of living matter. *J. Phys. Conf. Ser.* **2007**, *67*, 012006. [CrossRef]
31. Del Giudice, E.; Vitiello, G. Role of the electromagnetic field in the formation of domains in the process of symmetry-breaking phase transitions. *Phys. Rev. A* **2006**, *74*, 022105. [CrossRef]
32. Czerlinski, G.; Ryba, R. Coherence domains in living systems. *J. Vortex. Sci. Technol.* **2015**, *2*, 2. [CrossRef]
33. Montagnier, L.; Aissa, J.; Del Giudice, E.; Lavallee, C.; Tedeschi, A.; Vitiello, G. DNA waves and water. *J. Phys. Conf. Ser.* **2011**, *306*, 012007. [CrossRef]
34. Madl, P.; Del Giudice, E.; Voeikov, V.L.; Tedeschi, A.; Kolarž, P.; Gaisberger, M.; Hartl, A. Evidence of coherent dynamics in water droplets of waterfalls. *Water* **2013**, *5*, 57–68.
35. Carlon, H.R. Do clusters contribute to the infrared absorption spectrum of water vapor? *Infrared Phys.* **1979**, *19*, 549–557. [CrossRef]
36. Carlon, H.R. Variations in emission spectra from warm water fogs: Evidence for clusters in the vapor phase. *Infrared Phys.* **1979**, *19*, 49–64. [CrossRef]
37. Ryzhkina, I.S.; Murtazina, L.I.; Kiseleva, Y.V.; Konovalov, A.I. Supramolecular systems based on amphiphilic derivatives of biologically active phenols: Self-assembly and reactivity over a broad concentration range. *Doklady Phys. Chem.* **2009**, *428*, 201–205. [CrossRef]
38. Konovalov, A.I.; Ryzhkina, I.S. Formation of nanoassociates as a key to understanding of physicochemical and biological properties of highly dilute aqueous solutions. *Russ. Chem. Bull.* **2014**, *63*, 1–14. [CrossRef]
39. Yinnon, T.A.; Liu, Z.Q. Domains formation mediated by electromagnetic fields in very dilute aqueous solutions: 3. Quantum electrodynamic analyses of experimental data on solutions of weak electrolytes and non-electrolytes. *Water* **2015**, *7*, 48–69.
40. Messori, C.; Prinzera, S.V.; di Bardone, F.B. Deep into the water: Exploring the hydro-electromagnetic and quantum-electrodynamic properties of interfacial water in living systems. *Open Access Lib. J.* **2019**, *6*, 92815. [CrossRef]
41. Ivanov, V.E.; Usacheva, A.M.; Chernikov, A.V.; Bruskov, V.I.; Gudkov, S.V. Formation of long-lived reactive species of blood serum proteins induced by low-intensity irradiation of helium-neon laser and their involvement in the generation of reactive oxygen species. *J. Photochem. Photobiol. B Biol.* **2017**, *176*, 36–43. [CrossRef]
42. Ivanov, V.E.; Karp, O.E.; Bruskov, V.I.; Andreev, S.N.; Bunkin, N.F.; Gudkov, S.V. Formation of long-lived reactive products in blood serum under heat treatment and low-intensity laser irradiation, their role in hydrogen peroxide generation and DNA damage. *Indian J. Biochem. Biophys.* **2019**, *56*, 214–223.

43. Bunkin, N.F.; Lyakhov, G.A.; Shkirin, A.V.; Ignatiev, P.S.; Kobelev, A.V.; Penkov, N.V.; Fesenko, E.E. Mesodroplet heterogeneity of low-concentration aqueous solutions of polar organic compounds. *Phys. Wave Phenom.* **2019**, *27*, 91–101. [CrossRef]
44. Hwang, S.G.; Lee, H.S.; Lee, B.C.; Bahng, G. Effect of antioxidant water on the bioactivities of cells. *Int J. Cell Biol.* **2017**, *2017*, 1917239. [CrossRef]
45. Locke, B.R.; Shih, K.Y. Review of the methods to form hydrogen peroxide in electrical discharge plasma with liquid water. *Plasma Sources Sci. Technol.* **2011**, *20*, 034006. [CrossRef]

Article

Air Diffusion and Velocity Characteristics of Self-Aerated Developing Region in Flat Chute Flows

Liaochao Song [1], Jun Deng [2,*] and Wangru Wei [2]

[1] China Energy Dadu River Hydropower Development Co., Ltd., Chengdu 610065, China; songliaochao@sina.com
[2] State Key Laboratory of Hydraulics and Mountain River Engineering, Sichuan University, Chengdu 610065, China; wangru_wei@hotmail.com
* Correspondence: djhao2002@scu.edu.cn

Abstract: Self-aerated flows in flat chutes are encountered downstream of the bottom outlet, in spillways with a small slope and in storm waterways. In the present study, the development of self-aeration in flat chute flow is described and new experiments are performed in a long flat chute with a pressure outlet for different flow discharge rates. The distribution of air concentration, time mean velocity and velocity fluctuation in flow direction in the self-aerated developing region—where air bubbles do not diffuse next to the channel bottom—were measured and analyzed. The region of self-aeration from free surface was about 27.16% to 51.85% of the water depth in the present study. The analysis results revealed that the maximum distance of air bubble diffusion to the channel bottom increased with the development of self-aeration along the flow direction. This indicates that for flat chute flow, the process of air bubble diffusion from free surface to channel bottom was relatively long. Cross-section velocities increased along the flow direction in the self-aerated developing region, and this trend was much more remarkable in the area near water free surface. The velocity fluctuations in flow direction in cross-sections flattened and increased with the development of self-aerated flow. Higher velocity fluctuations in flow direction corresponded to the presence of much stronger turbulence, which enhanced air bubble diffusion from the water free surface to channel bottom along the flow direction.

Keywords: self-aeration; chute flow; air concentration; velocity; experimental study

Citation: Song, L.; Deng, J.; Wei, W. Air Diffusion and Velocity Characteristics of Self-Aerated Developing Region in Flat Chute Flows. *Water* **2021**, *13*, 840. https://doi.org/10.3390/w13060840

Academic Editor: Maksim Pakhomov

Received: 14 February 2021
Accepted: 10 March 2021
Published: 19 March 2021

1. Introduction

Free surface self-aeration is frequently observed in steep rivers, spillways and flat chutes, such as tunnel spillways, partially filled pipes and high-speed flow in open channels connecting to a pressure outlet in bottom outlet tunnels. The air entrainment induces a drastic change in the fluid characteristics. First, the amount of bulking caused by the entrained air makes the water depth higher than that of nonaerated water flow [1]. Aeration can eliminate or minimize cavitation damage caused by high-velocity flow in spillways, chutes and channels [2]. Studies have shown that the presence of air within the boundary layer can reduce the shear stress between the flow layers [3,4]. More recently, highly aerated flow has been recognized for its gas transfer characteristics with the transfer of atmospheric gases into the water and the volatilization of pollutants [5].

For an open channel flow in a flat chute with a pressure outlet, at the upstream end of the channel, a turbulent boundary layer which is generated by the channel bottom develops along the channel. The boundary layer thickness δ has been studied by many researchers [6–8]. The data of Schwarz and Cosart [9] suggested that $\delta/d \sim 0.18$ next to the intake, where d is the approaching flow depth. Chanson [10] showed that boundary layer thickness is $\delta \sim x^{0.8}$ on a smooth plate, where x is the streamwise distance from the outlet. Previous studies have shown that the position of the intersection (point A in Figure 1) between the water free surface and the outer edge of the turbulent boundary layer happens at a short distance from the intake. When the turbulence acting next to the free surface is great enough to overcome the surface tension effect and buoyancy, air bubble entrainment occurs. Chanson [11] stated that the normal turbulent velocities at the surface should be larger than 0.1 to 0.3 m/s for air bubble entrainment. For a high-velocity flow discharging into a flat chute, as in the experiments of Arreguin and Echavez [12], Anwar [13] indicated that the inception point of self-aeration (point B in Figure 1), where the location of the "white water" apparition appears, is closer to the upstream end than to the intersection of the boundary layer edge with the free surface.

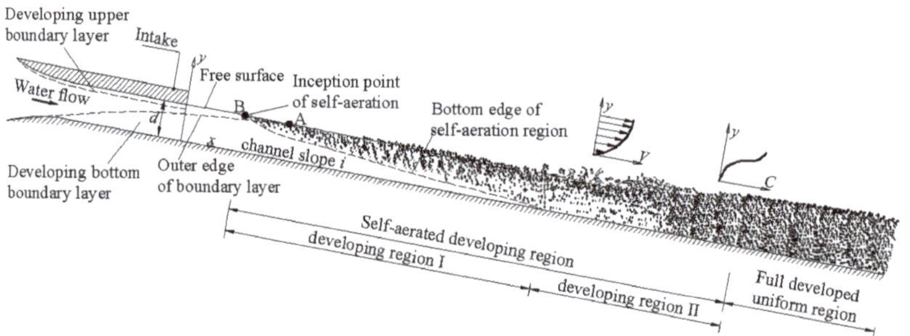

Figure 1. The region of self-aerated developing flow.

This means that the classical "boundary layer growth" theory for the steep slope trend, i.e., that the convergence of the inception point with the boundary layer and free surface intersection, does not apply to the inception of self-aeration in flat chutes. The flow separation and developing upper boundary layer need to be taken into consideration. The first part of the two-phase air–water flow is the self-aerated developing region, which forms before the flow gets to uniform equilibrium, and the second is the fully developed uniform region, where the turbulence diffusion is normal to the bottom and counterbalances exactly the buoyancy effect, and thus the air concentration is independent of the flow direction along the channel. In the self-aerated developing region, air bubble entrainment develops and diffuses to the maximum depth; this area along the channel is seen as developing region I. In this region, there are mainly three areas, that is, individual water drops in air, individual air bubbles in the water and clear water. For the area of individual water drops in the air, the most significant effects are on the water depth of aerated flow [14–16]. The transition section between individual water drops in the air and individual air bubbles is usually seen as the water depth of aerated flow increases. This section is also where the maximum gradient of air concentration from normal to the channel bottom appears [17]. Downstream of the developing region I, the air concentration distribution gradually varies along the channel, with more air bubbles entrained into the water flow and diffusing to the bottom. This distance in air–water flow can be recognized as developing region II.

A large amount of data about the self-aerated developing region II and the fully developed uniform region are available to predict the air–water flow properties of supercritical flows [18–20]. There is, however, little information on the self-aerated developing region I in flat chute flows, especially about the diffusing process of air bubbles to the bottom in developing region I (i.e., the area of individual air bubbles in water). In this paper, experimental data about the air concentration, time mean velocity and velocity fluctuation are obtained using a flat chute with a pressure outlet, and these characteristics of air–water flow are analyzed to further understand the process of self-aeration in high-velocity open channel flow.

2. Experimental Apparatus

The experiments were conducted in the state key lab of hydraulics and mountain river engineering, Sichuan University. The flume was 12 m long and 0.4 m wide with a

10 degree slope. The flume was made of planed synthetic glass boards (the roughness is 0.01 mm). Flow to the flume was fed through a smooth convergent nozzle (1.5 m long), as shown in Figure 2. The nozzle exit was 80 mm high. The flow discharge was delivered by a pump connected to a closed-circuit water supply system, and a thin-plate weir was set downstream to enable an accurate discharge adjustment. Air concentration and velocity measurements were performed on the channel centerline at regular intervals along the channel, and the first cross-section measurement was 2.7 m to the pressure outlet. For high-velocity air–water flow, an intrusive phase detection probe was the most reliable instrument for air–water property measurement [21]. The principle of electrical probes is the difference in electrical resistivity between air and water, and this can give accurate information on the local air–water fluctuations. In the present study, air bubble property and air–water velocity were recorded using the CQY—Z8a measurement instrument (made by Huaihe River Water Institution, Bengbu, China), a double-tip conductivity probe. The probe consisted of two identical tips with an internal concentric electrode made of platinum with a 0.04 mm diameter and an external stainless-steel electrode with a 0.7 mm diameter. Air–water velocity measurement was based on two successive detections of air–water interfaces by the two tips, and the computation was according to a cross-correlation technique between the two tip signals [22].

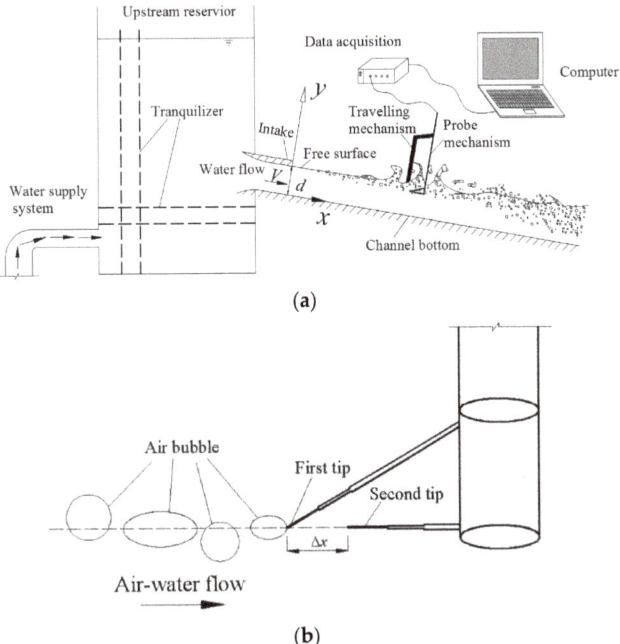

Figure 2. Sketch of experimental setup: (**a**) Schematic depiction of experimental configuration; (**b**) Sketch of the double-tip conductive probe.

The two tips were aligned in the flow direction, and the distance between the two tips was $\Delta x = 10$ mm. The cross-correlation function between the two tip signals was the maximum value of the cross-correlation coefficient R for the average time T_m taken for an "air bubble" interface to travel from the first tip to the second tip. The velocity was deduced from the time delay between the two tip signals and the two tip separation distance, Δx.

$$V = \Delta x / T_m, \tag{1}$$

This electronic system was designed with a response time of less than 10 μs. At each position, the two tip signals from the conductivity probe were recorded with a scan rate of 100 kHz per channel for a 5 s scan period.

Clear water velocities were measured with a Pitot tube (4 mm external diameter), and water velocity fluctuation in flow direction was recorded by a pressure sensor (CY200 made by Chengdu Test Electronic Information, Chengdu, China; comprehensive accuracy of 0.1%) connected to the Pitot tube. The vertical translation of the probe was controlled by an adjustment travelling mechanism and the error on the vertical position of the probe was less than 0.1 mm. The experiment was repeated until sufficient and reliable data were obtained at each measurement profile. The flow discharge rates, Q, were 0.18 m^3/s, 0.19 m^3/s, 0.20 m^3/s and 0.21 m^3/s, respectively, and Reynolds number was approximately within 3×10^5~4×10^5.

3. Experimental Results and Discussions

3.1. Air Concentration Distribution and Development

Distributions of air concentration and velocity were measured on the centerline (from x = 2.7~7.7 m). Typical results are plotted in Figure 3. Air concentration C is a function of y/d, where y is the normal distance to the channel bottom and d is the intake high. It is clear that in water flow, the air concentration increased continuously from the bottom to the free surface. At a certain cross-section, the air concentration increased much more rapidly near the water free surface, and in the self-aerated developing region, the distributions at the cross-sections are almost the same. The Froude number $Fr_0 = V/(gd)^{0.5}$ can be used as a flow condition for the present analysis, where V is the average flow velocity at the nozzle and g is the acceleration of gravity.

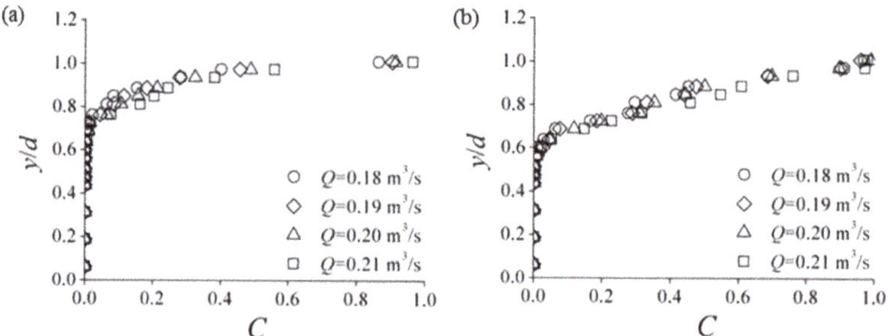

Figure 3. Comparison of air concentration distributions for different flow discharge rates: (**a**) x/d = 40; (**b**) x/d = 90.

In the present work, the air concentration distributions varied gradually, and the development of air bubble diffusion was not next to the bottom wall. Figure 4 shows the comparison between the present data and the two-dimensional diffusion model [10], defined as:

$$C = 1 - \tanh^2(K' - \frac{y'}{2 \times D'}), \quad (2)$$

where D' and K' are two constant parameters deduced from the mean air concentration of the cross-section. Although the distribution tendency of air concentration from the diffusion model basically fit the experimental air concentration profile, the difference in the bottom of the self-aeration region between the experimental data and the diffusion model is obvious.

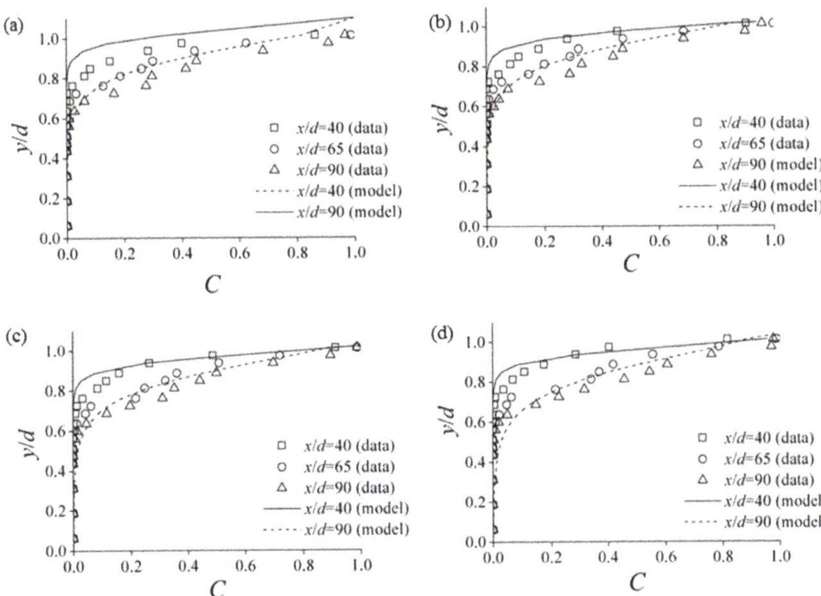

Figure 4. Air concentration distributions in cross-sections: (**a**) $Fr_0 = 6.353$; (**b**) $Fr_0 = 6.706$; (**c**) $Fr_0 = 7.059$; (**d**) $Fr_0 = 7.412$.

Figure 5 shows the development of the self-aeration region along the flow direction, compared with the results calculated by Equation (2) and Chanson's measured data [10]. The maximum normal distance y_{min} of the air bubble diffusing to the channel bottom in the present study was defined as the position where no air bubble signal was obtained by the probe. In Chanson's study, the development of air bubble diffusion at the channel bottom was relatively slower than that found in the present study. This is mainly because the channel slope of Chanson's study was much flatter than that of the present study. The theoretical diffusion model overestimates the rate of self-aeration development in flow cross-sections. For $Fr_0 = 6.353$, when the characteristic distance x/d increased from 33.75 to 96.25, the value of y_{min}/d varied from 0.725 to 0.563, while in the diffusion model it varied from 0.813 to 0.313. For $Fr_0 = 7.412$, when the characteristic distance x/d increased from 33.75 to 96.25, the value of y_{min}/d varied from 0.688 to 0.475, while in the diffusion model, it varied from 0.850 to 0.188. In the present study, when x/d increased from 33.75 to 96.25, the region of air bubble diffusion in the flow was about 27.16% to 51.85% of the water depth (the position at y direction, where the air concentration was larger than 0.9, was approximately the water depth in the developing region of self-aeration). In Chanson's results, the maximum difference of maximum normal distance of air bubble diffusion between experimental data and the diffusion model was at $x/d = 66.67$, whereas results from this study's experimental data and the diffusion model are $y_{\text{min}}/d = 0.942$ and $y_{\text{min}}/d = 0.455$, respectively. The development of air bubble entrainment in self-aerated flow is complicated, especially in the self-aerated developing region. The process is affected by incoming flow conditions, the solid wall, the channel slope, and the fact that the interaction between air bubbles and water is a three-dimensional process. A more accurate description of the air bubble entrainment is needed, while paying greater attention to the development process.

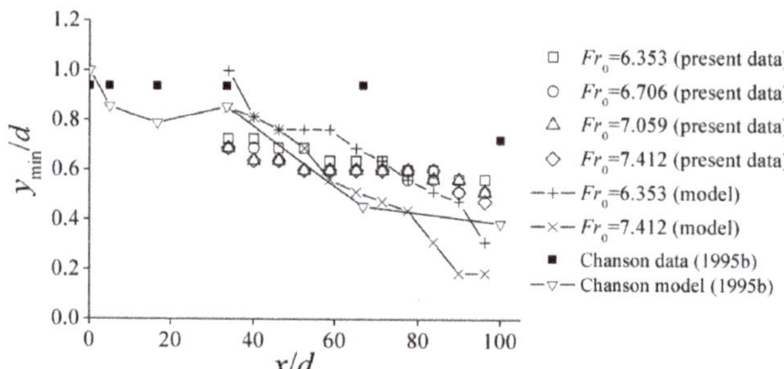

Figure 5. Comparison between experimental data and calculated results for the development of air bubble diffusion to the channel bottom.

3.2. Time Mean Velocity Distribution and Development

Time mean velocities in cross-sections were measured at various centerline positions from x = 2.7~7.7 m (i.e., x/d = 33.75~96.25) for different flow discharge rates. The velocity distributions at cross-sections for different flow discharge rates are plotted in Figure 6, where V_0 is the mean velocity, obtained from the ratio of Q to the square of intake (0.08 × 0.4 m).

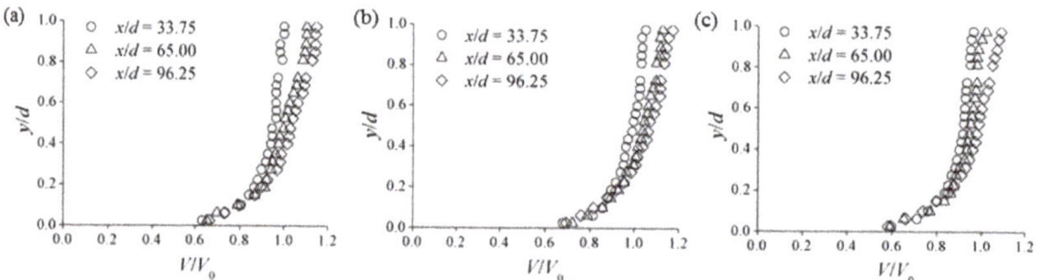

Figure 6. Velocity distributions in cross-sections along the flow direction: (**a**) Q = 0.18 m^3/s; (**b**) Q = 0.19 m^3/s; (**c**) Q = 0.21 m^3/s.

In the self-aerated developing region I, the cross-section velocities increased along the flow direction. This showed that the flow was non-uniform in self-aerated developing region I. The dimensionless velocity distributions are shown in Figure 7 and are compared with the logarithmic law for fully developed uniform flow. V_{mean} is the mean velocity of a certain cross-section. Considering a rough chute [23], the logarithmic law equations are:

$$\frac{V}{V_*} = 5.75\lg\frac{y}{\Delta} + 8.5 \quad (3)$$

where V_* is the friction velocity, obtained with the equations:

$$V_* = \sqrt{\frac{\lambda}{8}} \times V_{\text{mean}}, \quad (4)$$

$$\lambda = \frac{1}{\left[2\lg\left(3.7\frac{d}{\Delta}\right)\right]^2}, \quad (5)$$

where λ is the coefficient of friction resistance and Δ is the roughness (0.01 mm). For the cross-section near the intake (x/d = 33.75), the velocity distribution from experimental data is in good agreement with the logarithmic law. This is because the flow in intake is in the state of pressure flow, and it can be seen approximately as a uniform flow. For the cross-section near the intake in open channel flow, the air concentration was relatively low and the flow was still affected by the state of pressure uniform flow in the intake. This resulted in good agreement between the experimental data and the logarithmic law for uniform flow. With the development of flow downstream, self-aeration developed and the flow was in a state of acceleration under gravity, and this made the flow non-uniform.

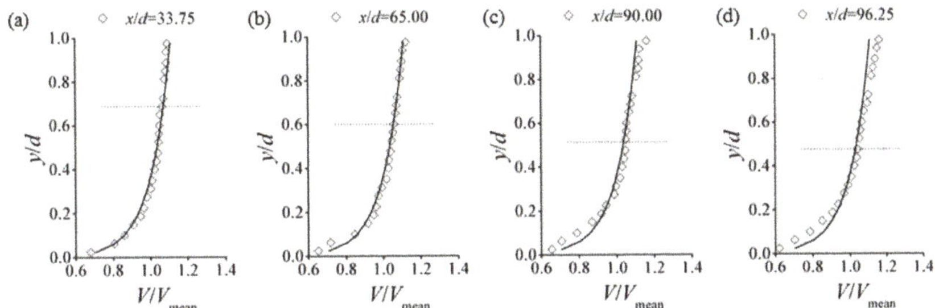

Figure 7. Mean time velocity profiles in the self-aerated developing region in chute flow for discharge rate $Q = 0.21\ m^3/s$: ◇, experimental data; solid line, the log law, Equations (3)–(5); dot-dashed line, the maximum distance of air bubble diffusion to the channel bottom at each cross-section.

For the cross-section far from intake ($x/d = 96.25$), the velocities near the free surface increased as the flow developed downstream, compared with the log-law results. Considering the aeration in flow, the viscosity and shearing stress in the flow were less than those in the non-aerated flow [15]. Thus, the air–water velocity will be comparatively greater than the clear-water velocity. In the self-aerated developing region, the effect of decreased resistance on velocity was developed with air bubble diffusion into the flow. With the development of aeration in flow, resistance decreased and velocity increases in the aeration region, while it lagged in the self-aerated developing region. For a certain cross-section, the position of velocity increase in the self-aerated region was closer to the free surface of flow in the self-aerated developing region than to the maximum distance of air bubble diffusion to the channel bottom (Figure 7). Figure 8 shows the near free surface distributions of time mean velocity, and the process of resistance decrease and velocity increase in the self-aerated developing region can be seen. On the other hand, according to the property of flow movement, the water acceleration process near the channel bottom was lagged compared with the water acceleration near the free surface because of the solid surface boundary. Thus, in the self-aerated developing region, velocities of the non-uniform flow near the channel bottom still developed and were relatively smaller compared with the velocity distribution obtained from the logarithmic law for uniform flow. The difference in velocity distribution at cross-sections indicated that for the self-aerated developing region in the flat chute flow, the velocity distribution deviated from the logarithmic law for the fully developed uniform flow.

3.3. Velocity Fluctuation and Development

For self-aerated flow in the open channel, flow turbulence had a significant effect on air bubble diffusion and self-aeration development in the flow. On the channel centerline, flow fluctuations in flow direction (x direction) were measured to analyze the flow turbulence. Considering the effect of the solid surface on the measurement setup, the flow fluctuations at the transverse direction (y direction) could not be measured accurately under present conditions. Based on Kolmogorov's hypothesis of local isotropy [24], at a sufficiently high Reynolds number, small-scale turbulent motions are statistically isotropic. In the present study, Reynolds number was within $3 \times 10^5 \sim 4 \times 10^5$, and this indicated that the flow fluctuations in flow direction (x direction) could basically reflect the flow turbulence qualitatively. At a certain cross-section, the measurement range was from $y/d = 0.025$ to

$y/d = 0.725$ as measured by the effects of air bubbles on the Pitot tube in the flow. The root mean square of pressure measured by transducer is given by:

$$\sigma = \sqrt{\frac{\sum_{k=1}^{n} (p_0 - p_k)^2}{n}}, \tag{6}$$

where p_k is the pressure recorded by the pressure transducer per signal, p_0 is the mean pressure during the scanning period at each measurement point and n is the total number of pressure signals ($n \geq 1$).

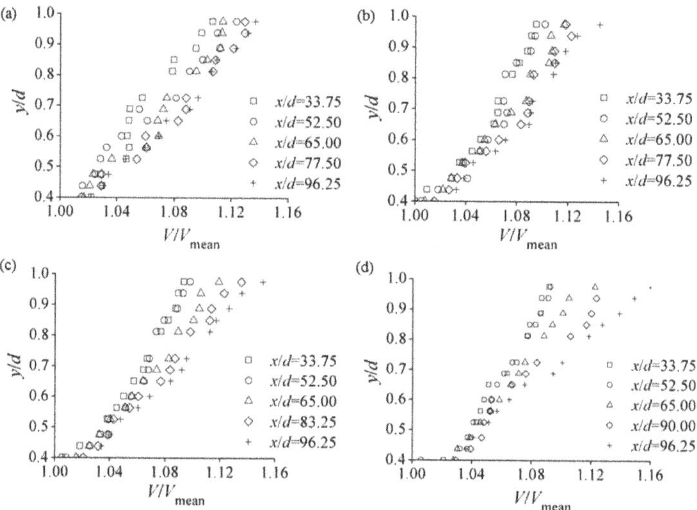

Figure 8. Near-free surface distribution of time mean velocity: (**a**) $Q = 0.18\ m^3/s$; (**b**) $Q = 0.19\ m^3/s$; (**c**) $Q = 0.20\ m^3/s$; (**d**) $Q = 0.21\ m^3/s$.

Figure 9 shows the dimensionless distributions of flow fluctuation σ/p_0 at different cross-sections in the self-aerated developing region. The first one or two values of σ/p_0 were relatively small, and this indicates that this area was affected the viscous sublayer where the flow type is mainly laminar. The viscous sublayer was caused by the solid wall of the channel bottom where the fluctuation was relatively low. With the increased distance from the channel bottom, the flow fluctuation at x direction increased rapidly, reached peak and then gradually fell. The value of σ/p_0 was relatively stable near the water free surface. The distributions of flow fluctuation in flow direction similarly followed the classical "boundary layer theory". The fluctuation in water flow developed from the channel bottom and subsequent transport to the water free surface. For a certain flow discharge rate, with increased distance to the nozzle intake ($x/d = 33.75$~96.25), the distribution of flow fluctuations in flow direction flattened, and the values of σ/p_0 near the water free surface increased. These changes reflected flow fluctuation at x direction properties in the self-aerated developing region in the flat chute flow.

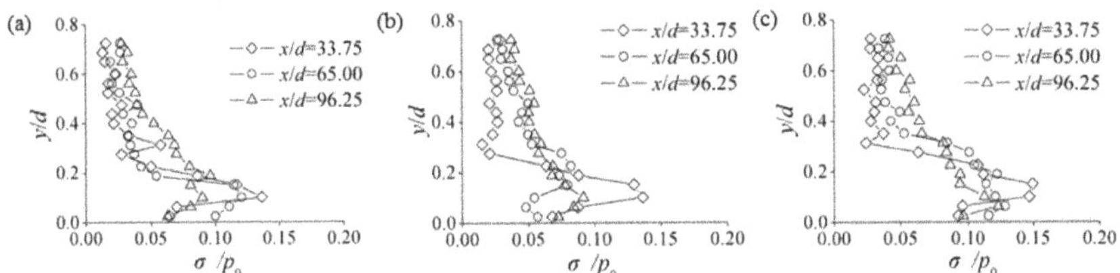

Figure 9. Dimensionless distributions of flow fluctuation at x direction in cross-sections: (**a**) $Q = 0.18\ m^3/s$; (**b**) $Q = 0.19\ m^3/s$; (**c**) $Q = 0.21\ m^3/s$.

Considering the ratio of the mean flow fluctuation at x direction of each cross-section σ_0 to the mean flow fluctuation at x direction of the high-velocity aerated chute flow σ_{mean} at the centerline, the ratio σ_0/σ_{mean} increased along the flow direction, as shown in Figure 10. The ratio σ_0/σ_{mean} showed the development process of flow fluctuation in the self-aerated developing region in the flat chute flow. With the self-aerated flat chute flow developing downstream, the fluctuation level increased. Similar trends were observed for the development of flow fluctuation at x direction in the self-aerated developing region in the present study. Considering that the flow fluctuations corresponded to the turbulence of flow, with the development of water flow downstream, the turbulence intensity increased. This increased the fluctuation enough to overcome the effects of buoyancy of air bubbles and to strengthen the air bubble diffusion inside the flow in the self-aerated developing region. Higher flow fluctuation corresponded to the presence of much stronger turbulence, which enhanced both the increased mean air concentration at the cross-section along the flow direction and the air bubble diffusion from the water free surface to the channel bottom.

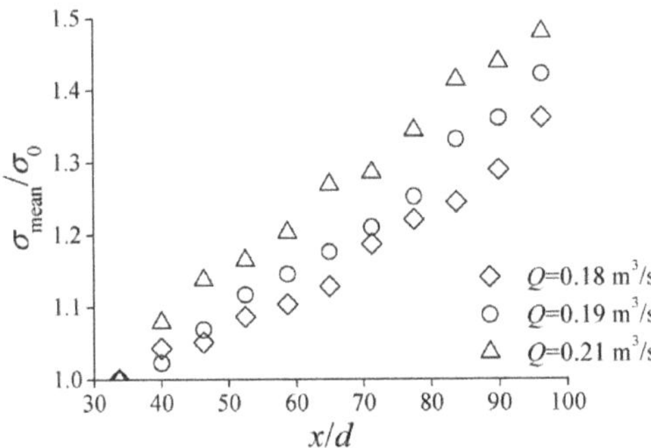

Figure 10. Dimensionless distributions of flow fluctuation at x direction along the flow direction.

4. Conclusions

New experiments were carried out to study the characteristics of the self-aerated developing region where the air bubbles do not diffuse next to the channel bottom in flat chute flows. With the flow developing downstream, air bubbles from self-aeration diffused to the channel bottom and the mean air concentration in cross-sections increased along the flow direction. The region of self-aeration from the free surface was about 27.16% to 51.85% of the water depth in the present study, and this indicated that for flat chute flow, the process of air bubble diffusion from free surface to channel bottom was relatively long. Time mean velocity and mean velocity fluctuation in flow direction in cross-section increased along the flow direction in the self-aerated developing region. In this area, the velocity fluctuation in flow direction in cross-sections flattened with the flow development downstream. In terms of turbulence, higher velocity fluctuation in flow direction corresponded to the presence of much stronger turbulence, which enhanced air bubble diffusion from the water free surface to the channel bottom and increased the mean air concentration in cross-section along the flow direction.

Author Contributions: This paper is a product of the joint efforts of the authors who worked together on the experimental model tests. J.D. has a scientific background in applied hydraulics, while L.S. and W.W. conducted the experimental data investigations. J.D. and L.S. tested the methods and wrote this paper. All authors have read and agreed to the published version of the manuscript.

Funding: Resources to cover the Article Processing Charge were provided by the National Natural Science Foundation of China (Grant No. 51939007) and Sichuan Science and Technology Program (Grant Nos. 2020YJ0320 and 2019JDTD0007).

Institutional Review Board Statement: Not applicable.

Informed Consent Statement: Not applicable.

Data Availability Statement: Not applicable.

Conflicts of Interest: The authors declare no conflict of interest.

References

1. Falvey, H.T. *Air-Water Flow in Hydraulic Structures*; USBR Engrg Monograph, No. 41: Denver, CO, USA, 1980.
2. Wood, I.R. *Air Entrainment in Free-Surface Flows*; IAHR Hydraulic Structures Design Manual No. 4, Hydraulic Design Considerations, A.A Balkema: Rotterdam, The Netherlands, 1991.
3. Wood, I.R. Uniform region of self-aerated flow. *J. Hydraul. Eng.* **1983**, *109*, 447–461. [CrossRef]
4. Chanson, H. Drag reduction in open channel flow by aeration and suspended load. *J. Hydraul. Res.* **1994**, *32*, 87–101. [CrossRef]
5. Chanson, H. Predicting oxygen content downstream of weirs, spillways and waterways. *Proc. ICE Water Marit. Energy* **1995**, *112*, 20–30. [CrossRef]
6. Schlichting, H. *Boundary Layer Theory*, 7 ed.; McGraw-Hill: New York, NY, USA, 1979.
7. Bauer, W.J. Turbulent boundary lauer on steep slopes. *Trans. Am. Soc. Civil. Eng.* **1954**, *119*, 1212–1233. [CrossRef]
8. Wang, J.Y. A study on computation of water depth in high velocity open channel flow. *J. Hydraul. Eng.* **1981**, *5*, 48–52. (In Chinese)
9. Schwarz, W.H.; Cosart, W.P. The two-dimensional wall-jet. *J. Fluid Mech.* **1961**, *10*, 481–495. [CrossRef]
10. Chanson, H. Air Bubble Entrainment in Free-Surface Turbulent Shear Flows. 1996. Available online: https://www.sciencedirect.com/book/9780121681104/air-bubble-entrainment-in-free-surface-turbulent-shear-flows?via=ihub=#book-description (accessed on 14 March 2021).
11. Chanson, H. Self-Aerated Flows on Chutes and Spillways. *J. Hydraul. Eng.* **1993**, *119*, 220–243. [CrossRef]
12. Arreguin, F.; Echavez, G. Natural Air Entrainment in High Velocity Flows. In Proceedings of the Conference on Advancements in Aerodynamics, Fluid Mechanics and Hydraulics, Minneapolis, MN, USA, 3–6 June 1986; Available online: https://cedb.asce.org/CEDBsearch/record.jsp?dockey=0048614 (accessed on 14 March 2021).
13. Anwar, H.O. Self-aerated flows on chutes and spillways-discussion. *J. Hydraul. Eng.* **1994**, *120*, 778–779. [CrossRef]
14. Morris, H.M. *Applied Hydraulics in Engineering*; Throndald Press Company: London, UK, 1963.
15. Wu, C.G. A research on self-aerated flow in open channels. *J. Hydroelectric Eng.* **1988**, *4*, 23–36. (In Chinese)
16. Afshar, N.R.; Asawa, G.L.; Raju, K.G.R. Air concentration distribution in self-aerated flow. *J. Hydraul. Res.* **1994**, *32*, 623–631. [CrossRef]
17. Straub, L.G.; Anderson, A.G. Experiments on Self-Aerated Flow in Open Channels. *J. Hydraul. Div.* **1958**, *84*, 1–35. [CrossRef]
18. Chanson, H. Uniform aerated chute flow-discussion. *J. Hydraul. Eng.* **1992**, *118*, 944–945. [CrossRef]
19. Yang, Y.Q.; Wu, C.G. A study on air concentration distribution of the developing zone in self-aerated open channel flows. *J. Chengdu Uni. Sci. Technol.* **1992**, *63*, 29–34. (In Chinese)
20. Bachalo, W. Experimental methods in multiphase flows. *Int. J. Multiph. Flow* **1994**, *20*, 261–295. [CrossRef]
21. Toombes, L.; Chanson, H. Surface waves and roughness in self-aerated supercritical flow. *Environ. Fluid Mech.* **2007**, *5*, 259–270. [CrossRef]
22. Chanson, H. Measuring air-water interface area in supercritical open channel flow. *Water Res.* **1997**, *31*, 1414–1420. [CrossRef]
23. Wu, C.G. *Hydraulics*; Higher Education Press: Beijing, China, 2003. (In Chinese)
24. Pope, S.B. Turbulent Flows. *Meas. Sci. Technol.* **2001**, *12*, 2020–2021. [CrossRef]

MDPI
St. Alban-Anlage 66
4052 Basel
Switzerland
Tel. +41 61 683 77 34
Fax +41 61 302 89 18
www.mdpi.com

Water Editorial Office
E-mail: water@mdpi.com
www.mdpi.com/journal/water

www.ingramcontent.com/pod-product-compliance
Lightning Source LLC
LaVergne TN
LVHW070702170726
843515LV00004B/462